Energy and Sustainability VIII

WIT*PRESS*

WIT Press publishes leading books in Science and Technology.
Visit our website for new and current list of titles.
www.witpress.com

WIT*eLibrary*

Home of the Transactions of the Wessex Institute.
Papers published in this volume are archived in the WIT eLibrary in volume 237
of WIT Transactions on Ecology and the Environment (ISSN 1743-3541). The WIT
eLibrary provides the international scientific community with immediate and permanent
access to individual papers presented at WIT conferences.
Visit the WIT eLibrary at www.witpress.com.

Eighth International Conference on Energy and Sustainability

ENERGY AND SUSTAINABILITY 2019

Conference Chairman

A. Tadeu
ITeCons, University of Coimbra, Portugal
Member of WIT Board of Directors

International Scientific Advisory Committee

Organised by

Wessex Institute, UK
ITeCons, University of Coimbra, Portugal

Sponsored by

WIT Transactions on Ecology and the Environment
International Journal of Energy Production and Management

WIT Transactions

Wessex Institute
Ashurst Lodge, Ashurst
Southampton SO40 7AA, UK

Senior Editors

Editorial Board

Energy and Sustainability VIII

Editor

A. Tadeu
ITeCons, University of Coimbra, Portugal
Member of WIT Board of Directors

WIT*PRESS* Southampton, Boston

Editor:

A. Tadeu
ITeCons, University of Coimbra, Portugal
Member of WIT Board of Directors

Published by
WIT Press
Ashurst Lodge, Ashurst, Southampton, SO40 7AA, UK
Tel: 44 (0) 238 029 3223; Fax: 44 (0) 238 029 2853
E-Mail: witpress@witpress.com
http://www.witpress.com

For USA, Canada and Mexico

Computational Mechanics International Inc
25 Bridge Street, Billerica, MA 01821, USA
Tel: 978 667 5841; Fax: 978 667 7582
E-Mail: infousa@witpress.com
http://www.witpress.com

British Library Cataloguing-in-Publication Data

A Catalogue record for this book is available
from the British Library

ISBN: 978-1-78466-345-2
eISBN: 978-1-78466-346-9
ISSN: 1746-448X (print)
ISSN: 1743-3541 (on-line)

The texts of the papers in this volume were set individually by the authors or under their supervision. Only minor corrections to the text may have been carried out by the publisher.

Preface

This Volume contains papers presented at the 8th International Conference on Energy and Sustainability held in Coimbra, Portugal, organised by Wessex Institute of Technology (WIT) and Itecons/UC.

The Conference Chairman was António Tadeu from Itecons, Full Professor at the University of Coimbra, and Member of the WIT Board of Directors.

The Meeting was sponsored by the International Journal for Energy Production and Management and WIT Transactions on Ecology and the Environment.

The Meeting follows a series of very successful conferences which started in 2007 in The New Forest, UK, home of the Wessex Institute. This was followed by meetings in Bologna (2009), Alicante (2011), Budapest (2013), Kuala Lumpur (2014), Medellin (2015), and Seville (2017).

For many years, as energy use led to increased production and distribution grew cheaper, efficiency and effects on the environment were not matters of concern.

As a result, we are now facing a shortage of resources and are forced to deal with the effects of climate change. Now, more than ever, regulators, scientists and industry stakeholders must pursue sustainability for the energy sector.

From production to distribution, technological advances are helping the transition towards cleaner energy supply. Following the growth of renewable energies, emphasis is also being put on the development of more sustainable energy storage. Likewise, legislators are putting new, stricter policies in place to enforce a more efficient use of energy, and their implementation presents new challenges. Even though energy efficiency is improving, it is becoming apparent that energy consumption must be drastically reduced if sustainable development is to be achieved, and if both climate change goals and societal goals are to be met. The concept of energy sufficiency is now attracting serious attention.

For sustainable energy goals to be met, it is crucial that researchers and professionals from a wide range of fields (engineering, environment, economics and social sciences) come together to address multidisciplinary challenges and discuss the technological advances being made.

The range of topics covered by the Conference involved collaboration between different subject areas in order to arrive at optimum solutions. They include Energy efficiency, Energy storage and distribution, Sustainable buildings, Energy policies, Biomass and biofuels, Environmental risk management, Ecosystems modelling and many others.

The book is now being distributed throughout the world in digital and paper formats and archived in the eLibrary of the Wessex Institute at www.witpress.com/elibrary. This fulfils one of the objectives of WIT, whose function is to disseminate knowledge throughout the international community.

The Editors are especially grateful to the late Carlos A. Brebbia for his commitment to research in this multi-disciplinary area.

Last but not least, the Editors would like to express their gratitude to the members of the International Scientific Committee for their invaluable help in selecting the papers, and to all the authors for their contributions.

António Tadeu
Coimbra, 2019

Contents

ENERGY TRANSITION: EMBODIED ENERGY IN CLEAN TECHNOLOGIES AND THE NEED FOR INTERNATIONAL REGULATORY PROACTIVITY

MAGNUS ABRAHAM-DUKUMA
Te Piringa – Faculty of Law, University of Waikato, New Zealand

ABSTRACT

The present study is an observation of the absence of policy focus and preparedness to address the greenhouse footprints and eventual climate change implications of some clean energy systems in the energy transition debate. Electric vehicles, wind turbines and photovoltaics constitute sample clean technologies under the study. A technical literature review is the approach for forming an appreciable understanding of the life cycle emissions of the sample clean energy technologies. Content and critical legal analysis are adopted for the examination of international and domestic legal regimes on clean energy sources, with India, China, the United States and the United Kingdom as domestic foci. The results of the study are the reality of supposed clean energy technologies as sources of greenhouse gas emissions, although considerably low compared to fossil fuel sources, and the apparent lack of international and domestic legal mechanisms to address the life cycle emissions of these technologies and the need for proactive policy actions.

Keywords: climate change, energy transition, clean energy systems, embodied energy, law.

1 INTRODUCTION

It has been observed that current emission pledges of countries are not enough to hold or reduce global average temperature well below 2°C [1]. Such an observation further dents the already battered footprints of conventional fossil fuels. Smart energy consumption and exploiting renewable energies are worthy energy transition pathway [2]. Specifically, the need to decarbonise energy-intensive industries by renewable energy to protect the environment may seem a proactive climate action [3]. Efforts by different countries towards climate change mitigation and transition to clean energy is commendable too [4]. However, as climate change is an unprecedented environmental problem, so does it raise novel legal challenges [5]. Irrespective of the laudable potential of renewable energy systems, research [6] shows that the major environmental problem, which they pose, is the embodied greenhouse emission traceable to the construction phase or raw materials used for manufacturing clean energy system components. Another way to conceptualise this is to consider the energy returned on energy invested (EROEI). This refers to the "ratio of the energy produced by an energy-converting device over its lifetime to the amount of primary energy required for the manufacture, transport, construction, operation and decommissioning of the device" [7]. The energy invested in this respect refers to the energy requirements for the manufacture of components, installation, maintenance and decommissioning of the energy system. This calls for a cautious evaluation of any utopian vision or ambition to build a total clean energy system against the odds of present reality [8]. This research elucidates this issue with an examination of electric vehicles (EVs), wind turbines (WTs), and photovoltaics (PVs), without prejudice to other clean energy sources worthy of consideration. Against this background, the study examines international and domestic legal frameworks for possible provisions bothering on life cycle emissions of these sample clean technologies.

www.witpress.com, ISSN 1743-3541 (on-line)
doi:10.2495/ESUS190011

2 GREEN TRANSITION AND EMBODIED ENERGY CONCERNS

2.1 Electric vehicles and GHG emissions

The widespread conceptualisation of EVs as a disruptive technology, climate change mitigation and decarbonising tool [9] sounds ambitious, but it is fraught with embodied greenhouse gas (GHG) emissions due to energy mix uncertainties, among other grey issues that merit examination [10]. Over 50% of the climate change effects of EVs are traceable to the manufacturing stage [11]. In contrast with internal combustion engine vehicles (ICEVs), research [12] shows that electric vehicles (EVs) account for higher emissions during manufacturing stage, generating 10 tons of carbon dioxide (CO2), while ICEVs generate 6 tons of CO2 at production phase. Another study [13] estimates the global warming potential of EV production to range between 87 to 95 gs of CO2. The high global warming impact of EVs at manufacturing stage is traceable to energy mix for battery production [14]. This pitches the manufacturing phase environmental and climate change impact of an EV above that of an ICEV (that is a conventional fossil fuel-powered vehicle) [15]. A 2017 research [16] reveals some technical details about EV manufacturing processes relevant to the present research. Firstly, the mineral mining requirements to produce a Li-ion battery for an EV are: spodumene, nickel sulphide ore, manganese carbonate. Secondly, the core material production phase requires aluminium, chromium steel, copper, and polypropylene. Thirdly, the pouch cell production requires cathode, anode, electrolyte, separator, and casing. In all these procedural manufacturing requirements, the cathode manufacturing accounts for the most climate change effect in the battery production phase, hence the need for improvement in the electricity mix for the EV manufacturing industry [16, p. 1259]. The transition to cleaner energy by means of electric vehicles can be more credible if accompanied by cleaner energy mix [17]. Although a 2018 study [18] observes a gradual improvement of the total life cycle GHG potential of EVs, it also acknowledges embodied energy traceable to the supply of electricity and manufacture of EV batteries.

2.2 Wind turbines and GHG emissions

Conceptualising wind turbines as perfectly clean energy source may be a myopic view void of consideration of possible greenhouse gas emissions which manufacturing phase of the life cycle accounts for [19]. The embodied energy of a wind turbine refers to the energy requirements for the manufacturing, construction, installation and maintenance phases. When considering multiple wind turbines for a wind farm, other materials and components such as wiring, grid connection, transformers and access roads may also account for embodied energy of the wind farm [19]. Research [20] shows that wind turbine energy system is not entirely clean as different phases of the life cycle of a wind turbine are dependent on fossil fuel, therefore prone to carbon emissions. The installation phase requires the use of heavy-duty machinery to mount the various components. The required equipment/machines run on fossil fuel. Decommissioning phase also uses fossil fuel to power cranes and other heavy-duty equipment for dismantling and transportation to incineration or landfill sites. This process is also an emission factor [20]. In another study [21] to estimate the embodied carbon emissions from 14 wind turbines with rated powers between 50 Kw and 3.4 Mw, it was found that the materials and machinery for production ranged between 2,139–2,142 tons of CO2 for each turbine with steel, aluminium, polyester and copper accounting for higher emissions. The study notes that only accounting for the operational emissions of a wind turbine misrepresents the environmental impact of its lifecycle. In comparison with fossil fuel, wind energy is a more environmentally friendly energy source, but also has social and environmental footprints with climate change implications [22]. There are projections for technology improvement opportunities (TIOs) in the manufacturing materials and processes for the wind farm industry, with particular respect to an onshore wind farm [23]. Nevertheless, whether the context is for an onshore or an offshore wind farm, it is necessary to pursue a better understanding of ways to enable a cleaner energy transition devoid of (or with very infinitesimal) GHG footprints like conventional fossil fuels [24].

2.3 Photovoltaic system and GHG emissions

Photovoltaic (PV) or solar energy conversion is one of the renewable energy systems in the energy transition with great potentials to mitigate GHG emissions [25]. While it may seem clean and portends environmental and economic advantages in comparison with conventional fossil fuels [26], it is also fraught with emission concerns traceable to the use of fossil fuel sources as process energy [27]. Specifically, the emissions associated with the PV system come from the entire energy requirements for extraction of raw materials, purification, panel production and complete module installation [28]. Its generation system consists of cells, electrical and mechanical connections, and mounting; while its production materials include "solar cells of multi-crystalline-silicon (mc-Si), single-crystalline-silicon (Sc-Si), amorphous-silicon (a-Si), cadmium-telluride (CdTe), copper-indium-diselenide (CIS) and of other thin layer materials" [28]. The primary production phase energy input for PVs comprise mainly non-renewable energy sources: oil and natural gas for electricity and steam [29]. The batteries and panels embody considerable volumes of energy invested at production phase [30]. This electricity mix usually contributes 80% to 90% of the total production energy requirement and accounts for GHG emissions of photovoltaics [31]. The extent of fossil fuel invested at the manufacturing stage usually determines emission intensity of the PV system [32]. Earlier studies by Alsema [33], Bankier and Gale [34] estimate the embodied energy for mc-Si module comprising frame, supports and inverter as 5400 MJ/m2; 6900 MJ/m2 for Sc-Si; and 2400 MJ/m2 for Thin Film. Much of this embodied energy is primarily traceable to production of the manganese, cobalt and copper contained in the PV battery [35].

3 INTERNATIONAL LEGAL CONSIDERATIONS

International legal framework for clean energy appears to be fragmented, comprising different direct and indirect legal mechanisms [36]. The International Renewable Agency (IRENA) Statute [37], the International Energy Charter (IEC) [38] and the Paris Climate Change Agreement (PCCA) [39] are vital legal mechanisms with relevant provisions on energy transition. The promotion of widespread adoption of renewable energy sources forms the key objective of the International Renewable Energy Agency (IRENA) [37]. Similarly, the use of renewable energy sources and clean technologies are among the energy efficiency and environmental protection mechanisms provided for by the IEC [40]. The PCCA clearly acknowledges the deployment of renewable energy as a means of averting the environmental impacts of conventional fossil fuels [39, preamble]. These provisions by the different highlighted international legal instruments are laudable for the energy transition debate, with potential to foster a paradigm shift to a green energy economy globally [41]. More so, international policy on energy and renewables has become a powerful normative force for global energy policymaking throughout the years [42]. However, a keen reading of the provisions referred to above reveals that they make neither provisions on nor consideration of energy investment and the embodied energy requirements of widely acknowledged clean energy systems. It is either there is an apparent ignorance or a blatant disregard and undermining of the carbon footprints of supposed clean energy systems by the international legal order. If any of these possibilities is the case, an international legal consciousness of this issue becomes necessary. As we make relentless moves towards energy transition for a better environment, there is also need for comprehensive policy regimes to support and address multi-faceted issues and sustainability concerns associated with emerging energy technologies, at the international and domestic levels [43]. International organisations like the ones highlighted above, with their policy regimes, are well suited to engender consensus on the scientific and policy underpinnings of global challenges [44]. Such consensus can have a positive impact in raising awareness on the issue of embodied energy argued in this paper.

4 DOMESTIC LEGAL CONSIDERATIONS

China, the United States (US), India and the United Kingdom (UK) stand as top renewables growth leaders [45]. While China, the US and India stand as the top-three renewables growth leaders globally [45], the UK also stands as a leading renewables giant [46], with a leading position for power generation from offshore wind, generating a total of 6.8 gigawatts (GW) at the end of 2017, ahead of Germany (5.4 GW), China (2.8 GW), Denmark (1.3 GW) and Netherlands (1.1 GW) [46, p. 133]. According to the Climate Policy Initiative (CPI) and IRENA, these countries have also made massive financial investments in renewables [47]. Additionally, there is a positive paradigm shift in environmental agenda in China and India, two countries notorious for high carbon emissions [48]. This trend is not limited to China and India, but extends to the BRICS nations – Brazil, Russia, India, China and South Africa [49]. It is possible that there are other jurisdictions with well-advanced clean energy systems. In fact, Sweden, Costa Rica, Nicaragua, Scotland, Germany, Uruguay, Denmark, Morocco and Kenya are making massive investments in clean energy systems [50]. However, the paper limits its examination to China, the US, India and the UK.

4.1 China

The energy legal regime of China makes ample provisions for the use of renewable energy sources in the transition from fossil fuels with no consideration of greenhouse gas emissions footprints of clean energy systems. Particularly, the Chinese Energy Conservation Law [51] provides for state development and utilization of renewable energy sources like solar energy, wind energy among others [51, arts 4 and 11]. There are statutory [51, art 30], and policy supports for financing massive investments, promoting and conducting research and development and use of EVs in place of ICEVs [52]. A latter renewable energy law also makes provisions for promotions, adaptation and incentivizing the commercial development of clean energy sources [53]. Pursuant to the Chinese Renewable Energy Law [54], there exists a regulatory guidance (the Circular) for the industrial development of renewables [54, art 9]. The Circular makes provisions on specifications of different components of renewable energy systems but is silent about the energy invested in the process of manufacturing components, production and eventual installation of the systems. Rather, policy focus is on research and industrial development of the supposed clean energy alternatives to fossil fuel sources. In response to the encouraging regulatory and policy environment, there has been a massive development of renewable energy sources such as wind [55], EVs [56] and photovoltaics [57] in China. However, the regulatory regime on renewables is silent on the climate change implications of energy requirements for the manufacture and installation of the supposed clean energy systems in China. There is room for improvement in the energy conservation laws of China in this respect [58].

4.2 India

India's renewable energy drive hinges on national desire to divest from significant reliance on fossil fuels, achieving energy independence and climate change mitigation [59]. Years before the emergence of the PCCA in 2015 [39], India already had functional policy instruments for climate change mitigation and renewable energy development. Few are relevant to the present discourse. The Energy Conservation Act [60] provides for energy efficiency [60, s 3], but there is no provision for renewable energy development. The Integrated Energy Policy (IEP) [61], National Action Plan on Climate Change (NAPCC) [62] and the Energy Sector Plan [63] all provide for promoting energy efficiency, with emphasis on renewable energy sources. The different policy mechanisms make provisions for incorporating wind and photovoltaics into India's national energy mix. In fact, the NAPCC was India's national policy response to the 2009 Copenhagen Summit [64]. It is true that the Indian State navigated through the multi-faceted issues of economic growth, potential energy crisis and climate change [65]. Policy focus however was in meeting its surging energy demand [66] and mitigating climate change implications of energy use. These are good objectives and pragmatic solutions for energy security and sustainable development [67]. The different policies may have incentivised an appreciable emerging clean energy market [68]. Other regulatory instruments like the Electricity Act [69] and the Tariff Policy [70] may have also contributed to the development of renewables in India because of their provisions on energy efficiency and renewable energy [71]. Renewable energy has enabled promising energy security and sustainable energy architecture in India [72], but policy instruments seem to be blind to the scientific literature regarding the embodied energy of renewables. This is not to undermine commendable regulatory strides and policy advances made towards climate change mitigation by incentivising investments in and deployment of renewable energy sources in India. The intention is to draw attention to a subtle issue, which the stern eyes of the law should be wary of and possibly address.

4.3 The United States

The United States' policy on air pollution control and climate change has gone through a chequered history crisscrossing different legislative enactments [73]. From the mid-1990s until 2018, there have been over 100 policy and statutory enactments on air quality, energy and renewables in the US [74]. From the premier statutory enactment (Air Pollution Control Act 1955) [75] to the major Clean Air Act Amendments of 1990 [76] federal policy has been toward ensuring ambient air quality and mitigating climate change through renewables. However, the examination in this paper is limited to four major policy instruments, being the Energy Policy Act [77], Energy Independence and Security Act [78], American Recovery and Reinvestment Act [79] and the Clean Air Act (As Amended) [80]. The Recovery and Investment Act makes provision for deployment and use and of renewables [81]. It also provides for tax incentives for investment in this respect [81, s. 1101]. However, no provision or related provision addresses the climate change implications of renewables, especially at manufacturing phase. The Energy Policy Act [77] makes provisions for incentivising renewable energy production and increased use of renewables, including solar thermal energy, wind energy, among others. The focus seems to be on optimising renewables for energy security, conservation and efficiency. There is no provision or recognition of the energy invested into manufacturing the components of renewables, especially the energy requirement at the manufacturing phases of wind turbines and solar panels. The Energy Independence and Security Act [82] also makes salient provisions for grants [82, ss. 202,

207], university-based research and development program [82, s. 234], renewable fuel infrastructure development to retailers and wholesale dealers [82, 244] geared towards incentivising the development and use of renewables. Again, there is little regard for the environmental or climate implications of manufacturing phase greenhouse gas emissions of clean energy systems. The Clean Air Act makes very similar provisions for renewable fuel program [82, s. 211 (o)] like the Energy Independence and Security Act. The arguments advanced in respect of the Energy Independence and Security Act apply in equal measure to the Clean Air Act. However, it is important to add that the Regulator (being the Administrator of the Environmental Protection Agency) can optimise the regulatory powers conferred on him by the Clean Air Act to promulgate subsidiary legislation [82, s. 211) (v) (2)] to mitigate the embodied energy implications of renewables, as part of the bigger or broader climate change goal.

4.4 The United Kingdom

Climate change and energy security constitute key drivers of the UK's energy transition [83]. Its transition pathways encompass energy governance and technological choices for low-carbon energy systems [84]. Specifically, its legislative/policy portfolio or framework on climate change and renewables comprise multiple legislative enactments and policies such as the Energy Act 2013, Finance Act 2011, Feed-in Tariffs for Renewable Electricity 2014, Carbon Reduction Commitment Energy Efficiency Scheme 2014, Climate Change Act 2008, National Adaptation Programme 2013, the UK Renewable Energy Roadmap and few others [85]. There are also multiple policies (repealed and) in force which directly and indirectly constitute key elements of its clean energy policy framework [86]. However, this paper limits its examination to the Energy Act 2013, the UK Renewable Energy Roadmap 2011, Climate Change Act 2008, and the European Union Renewable Energy Directive 2009, with more focus on the latter two, as they constitute the primary regulatory framework for renewables in the UK [87]. The Climate Change Act [88] is arguably the first statutory mechanism globally (and consequently makes UK the first country) to set legally binding targets for emissions reduction and addressing climate change with its attendant consequences [89]. With its key objectives of improving carbon management, addressing climate change and projecting the UK as a strong global climate action leader, it provides for a short-term 26% emissions reduction below the 1990 levels [88, s. 5] and a long-term 80% emissions reduction lower than the 1990 baseline by 2050 [88, s. 1]. Consequently, it makes provisions for use of renewable fuels for transportation as a way of combating the climate change implications of fossil fuels.[88, s. 78] The European Union (EU) Renewable Energy Directive [90] requires the UK to source 15% of its national energy consumption from renewables by 2020 (and varied targets for other EU Member-States) [90, arts 3(1), 4(1) and ANNEX 1]. Pursuant to this provision, the UK Renewables Policy Framework for Action pursues financial support for investment in renewables, unblocking barriers to delivery, and development of emerging technologies – wind, solar photovoltaics, wave, hydro-electric, etc. [91]. These sentiments also form the directive principles and objectives of the UK's Renewable Energy Roadmap, as projected by the UK Department of Energy and Climate Change (DECC) [92]. Similar to the National Renewable Energy Action Plan (NREAP), the Roadmap identifies investment and deployment of eight clean technologies – onshore wind, offshore wind, renewable transport, marine energy, biomass electricity, biomass heat, ground source and air source heat pumps – policy/action pathways for achieving over 90% of UK's renewable energy goal for 2020 in a cost-effective way [93], [5]–[6]. The Energy Act [93] makes statutory provisions to incentivise emissions reduction in the UK energy mix. We find specific provisions in its recognition of the decarbonisation target set out in the EU Renewables Directive [93, s. 5(2)(e) and (3)]. More directly, Contracts for Difference (CFD), aimed at decarbonising electricity generation, capacity market and related functions constitute the cardinal features of the electricity market reform contemplated by the Act [93, s. 6]. Renewables Obligations (RO) place statutory duty on electricity companies to source increasing percentage of electricity generation from renewables [93, s. 32]. These legislative and policy incentives in the foregoing no doubt represent a good regulatory framework for achieving emissions reduction and addressing climate change at a national level [94]. The Climate Change Act in particular has the status of climate constitutionalism, the first of its kind globally [95]. The entire policy mix has been relatively successful in incentivising increased investment in renewables for the purposes of emissions reduction and less reliance on fossil fuels for energy needs [96], even if there is need for reduced renewables generating costs [96]. Further evidence in this respect is UK's position as an offshore wind energy leader [46], and one of the top 10 renewables financing countries [47]. Additionally, there are quite commendable low carbon, renewable and energy efficiency projects across the over 400 local authorities in the UK [97]. However, a careful examination reveals that there seems to be no provision or coverage of embodied energy concerns of renewable energy technologies in the entire policy mix. Policy focus seems to be on emissions reduction from fossil fuel related energy generation and distribution companies. It is also good to think about the implicit or explicit environmental implications of renewables to avoid passing them off as perfectly or completely clean energy systems.

As energy technologies are evolving, there is need for a coordinated policy approach for policy support and the right regulatory framework. It is also important that policy measures at international and domestic fronts keep pace with technological advancement by legislative reviews of existing laws or entirely new policy measures to address seeming environmental and multi-facetted concerns associated with technologies not envisaged or contemplated by the current state of law. The law should be sensitive to innovation and technological advancements and reinvent itself accordingly [98]. At the domestic level, as the framer of innovation [99], the state has a role to play in addressing embodied energy concerns through innovative regulations. No doubt, there has been a growth of policy measures to incentivise renewable projects across diverse jurisdictions due to their clean energy potentials [100]. Irrespective of these, effective regulation is still necessary for addressing likely environmental and safety concerns [101].

5 CONCLUSIONS

From a complete life cycle perspective, touted clean energy systems such as wind turbines, electric vehicles and photovoltaics are not entirely clean. They do not totally obviate the environmental concerns associated with conventional fossil fuels. Thus, rather than representing them as entirely clean energy alternatives, it may be safer to assert that they imply lower global warming potentials in comparison with conventional fossil fuels. At the realm of international law, there is lack of policy provision or initiative to address the climate change implications of energy investment into these supposed clean technologies. The situation is similar in leading renewables jurisdictions such as China, India, the US and the UK, irrespective of their emerging legal frameworks on renewables. As innovation snowballing takes centre stage in the energy transition process, the law should also reinvent itself internationally and domestically and be proactive in envisaging and addressing likely or unanticipated challenges like embodied energy.

ACKNOWLEDGEMENTS

The paper was written during the course of a doctoral research project funded by the University of Waikato, Hamilton, New Zealand and was supported by the Section on Energy, Environment, Natural Resources and Infrastructure Law (SEERIL) of the International Bar Association (IBA).

REFERENCES

[1] Peters G.P. et al., Key indicators to track current progress and future ambition of the Paris Agreement. *Nature Climate Change,* **7**, pp. 118–122, 2017.

[2] Ngo, C. & Natowitz, J.B., *Our Energy Future: Resources, Alternatives and the Environment,* 2nd edn. Wiley: New Jersey, pp. 661, 2016.

[3] Ahman, M., Nilsson, L.J. & Johansson, B., Global climate policy and deep decarbonisation of energy-intensive industries. *Climate Policy,* **17**(5), pp. 634–649, 2017.

[4] Obama, B., The irreversible momentum of clean energy. *Policy Forum,* **355**(6321), pp. 126–129, 2017.

[5] Farber, D.A., Climate policy and the United States system of divided powers: Dealing with carbon leakage and regulatory linkage. *Transnational Environmental Law,* **3**(1), pp. 31–55, 2014.

[6] Gallagher, J. et al., Inventory compilation for renewable energy systems: The pitfalls of materiality thresholds and priority impact categories using hydropower case studies. *International Journal of Life Cycle Assessment,* **20**(12), pp. 1701–1707, 2015.

[7] Lloyd, B. & Forest, A.S., The transition to renewables: Can PV provide an answer to the peak oil and climate change challenges? *Energy Policy,* **38**(11), pp. 7378–7394, 2010.

[8] Clark, C.T.M. et al., Evaluation of a proposal for reliable lo-cost grid power with 100% wind, water and solar. *Proceedings of the National Academy of Sciences of the United States of America,* **114**(26), pp. 6722–6727, 2017.

[9] Muraya, N.K. & Capehart, B.L., Are EVs the digital equivalent of HDTV? *Energy Engineering,* **122**(6), pp. 11–32, 2015.

[10] Jochem, P., Babrowski, S. & Fichner, W., Assessing CO_2 emissions of electric vehicles in Germany in 2030. *Transportation Research Part A: Policy and Practice,* **78**, pp. 68–83, 2015.

[11] Zackrisson, M., Avellan, L. & Orlenius, J., Life cycle assessment of lithium-ion batteries for plug-in hybrid electric vehicles: Critical issues. *Journal of Cleaner Production,* **18**(15), pp. 1519–1529, 2010.

[12] Requia, W.J., Carbon dioxide emissions of plug-in hybrid electric vehicles: A life-cycle analysis in eight Canadian cities. *Renewable and Sustainable Energy Reviews,* **78**, pp. 1390–1396, 2017.

[13] Hawkins, T.R. et al., Comparative environmental life cycle assessment of conventional and electric vehicles. *Journal of Industrial Ecology,* **17**(1), pp. 53–64, 2012.

[14] Sanfélix, J. et al., Environmental performance of advanced hybrid energy storage systems for electric vehicle applications. *Applied Energy,* **137**, pp. 925–930, 2015.

[15] Casals, L.C. et al., Sustainability analysis of the electric vehicle use in Europe for CO_2 emissions reduction. *Journal of Clean Production,* **127**, pp. 425–437, 2016.

[16] Wang, Y., Quantifying the environmental impact of a li-rich high-capacity cathode material in electric vehicles via life cycle assessment. *Environmental Science Pollution Research,* **24**(2), pp. 1251–1260, 2017.

[17] Miotti, M., Hofer, J. & Bauer, C., Integrated environmental and economic assessment of current and future fuel cell vehicles. *International Journal of Life Cycle Assessment,* **22**(1), pp. 94–110, 2017.
[18] Wu, Z. et al., Life cycle greenhouse gas emission reduction potential of battery electric vehicle. *Journal of Cleaner Production,* **190**, pp. 462–470, 2018.
[19] Savino, M.M. et al., A new model for environmental and economic evaluation of renewable energy systems: The case of wind turbines. *Applied Energy,* **189**, pp. 739–752, 2017.
[20] Martin, E., The wind turbine: energy, design life-cycle. www.designlife-cycle.com. Accessed on: 5 Jan. 2019.
[21] Smoucha, E.A. et al., Life cycle analysis of the embodied carbon emissions from 14 wind turbines with rated powers between 50 Kw and 3.4 Mw. *Journal of Fundamentals of Renewable Energy and Applications,* **6**(4)**,** pp. 1, 2016.
[22] Kaldellis, J.K. et al. Environmental and social footprint of offshore wind energy: comparison with onshore counterpart. *Renewable Energy,* **92**, pp. 543–556, 2016.
[23] Ozoemena, M., Cheung, W.M. & Hasan, R., Comparative LCA of technology improvement opportunities for a 1.5-MW wind turbine in the context of an offshore wind farm. *Clean Technologies and Environmental Policy,* **20**(1), pp. 173–190, 2018.
[24] Duic, N., Is the success of clean energy guaranteed? *Clean Technology Environment Policy,* **17**(8)**,** pp. 2093–2100, 2015.
[25] Nikolić, D. et al., Optimization of photovoltaics panels area at Serbian zero-net energy building. *Journal of Renewable and Sustainable Energy,* **5**(4), pp. 041819-1, 2013.
[26] Ardente, F. et al., life cycle assessment of solar technologies. *Sustainability Assessment of Renewables-Bases Products: Methods and Case Studies,* eds J. Dewulf, S.D. Meester & R.A.F, Alvarenga, Wiley: West Sussex, pp. 241, 2016.
[27] Azzopardi, B. and Mutale, J., Life cycle analysis for future photovoltaic systems using hybrid solar cells. *Renewable and Sustainable Energy Reviews,* **14**, p. 1130, 2010.
[28] Burg, B.R. et al., Effects of radiative forcing of building integrated photovoltaic systems in different urban climates. *Solar Energy,* **147**, pp. 399–405, 2017.
[29] Fu, Y., Liu, X. & Yuan Z., Life cycle assessment of multi-crystalline photovoltaic (PV) systems in China. *Journal of Cleaner Production,* **86**, pp. 180–190, 2015.
[30] Mason, I.G. & Miller, A.J.V., Energetic and economic optimisation of islanded household-scale photovoltaic-plus-battery systems. *Renewable Energy,* **96**, pp. 559–573, 2016.
[31] Wong, J.H., Royapoor, M. and Chan, C.W., Review of life cycle analysis and embodied energy requirements of single-crystalline and multi-crystalline silicon photovoltaic systems. *Renewable and Sustainable Energy Reviews,* **58**, pp. 608–618, 2016.
[32] Fthenakis, V., Life cycle assessment of photovoltaics. *Photovoltaic Solar Energy,* eds R. Reinders et al., Wiley: West Sussex, pp. 646, 2017.
[33] Alsema, E.A., Energy pay-back time and CO_2 emissions of PV systems. *Progress in Photovoltaics: Research and Applications,* **8**(1), pp. 17, 2010.
[34] Bankier, C. & Gale S., Energy payback of roof mounted photovoltaic cells. *Energy Bulletin,* pp. 5, 2006.
[35] Bilich, A. et al., Life cycle assessment of solar photovoltaic microgrid systems in off-grid communities. *Environmental Science and Technology,* **51**(2), pp. 1043–1052, 2017.

[36] Am, M.W. & Drake, L., International law and the renewable energy sector. *The Oxford Handbook of International Climate Change Law,* eds C.P. Carlarne., K.R. Gray & Tarasofsky, OUP, pp. 357, 2016.
[37] The Statute of IRENA signed in Bonn on 26 January 2009.
[38] The International Energy Charter, adopted 20 May 2015 and entered into force 4 November, 2016.
[39] The Paris Agreement under the United Nations Framework Convention on Climate Change, adopted 12 December 2015 and entered into force 4 November 2016.
[40] IEC, Title 1(3), title II art 7.
[41] Gehring, M.W., Legal transition to the green economy. *McGill International Journal of Sustainable Development Law and Policy,* **12**, pp. 135–173, 2016.
[42] Bruce, S., International law and renewable energy: Facilitating sustainable energy for all? *Melbourne Journal of International Law,* **14**(1), pp. 1–36, 2013.
[43] Bradbrook, A.J., Creating law for next generation energy technologies. *George Washington Journal of Energy and Environmental Law,* **2**, pp. 17–38, 2011.
[44] Meyer, T., Global public goods, governance risk, and international energy. *Duke Journal of Comparative and International Law,* **22**, pp. 319–348, 2012.
[45] IEA, *Renewables 2017: Analysis and Forecasts to 2022,* IEA, 2017.
[46] REN21, *Renewables 2018: Global Status Report* (Renewable Energy Policy Network for the 21st Century 2018) at 71, at 90 and 109.
[47] CPI and IRENA, *Global Landscape of Renewable Energy Finance 2018,* IRENA, Abu Dhabi, pp. 27, 2018.
[48] Jayaram, D., A shift in the agenda for China and India: Geopolitical implications for future climate governance. *Carbon and Climate Law Review,* **9**(3), pp. 219–230, 2015.
[49] Sahu, M.K., Energy revolution under the BRICS nations. *BRICS Law Journal,* **3**(1), pp. 34–42, 2016.
[50] The Climate Reality Project, follow the leader: How 11 countries are shifting to renewable energy; The Climate Reality Project. www.climaterealityproject.org/blog/follow-leader-how-11-countries-are-shifting-renewable-energy. Accessed on: 4 Dec. 2018.
[51] Energy Conservation Law of the People's Republic of China, No. 90, 1997.
[52] Zhang, X. et al., Review of electric vehicle policies in China: Content summary and effect analysis. *Renewable and Sustainable Energy Reviews,* **70**, pp. 698–714, 2017.
[53] Renewable Energy Law of the People's Republic of China 2005, arts 1 and 2.
[54] Circular of the National Development and Reform Commission on Printing and Distributing Ctalog for the Guidance of the Industrial Development of Renewable Energy (Fa Gai Energy 2005 No. 2517).
[55] Yang, J. et al., The life-cycle energy and environmental emissions of a typical offshore wind farm in China. *Journal of Cleaner Production,* **180**, pp. 316–324, 2018.
[56] Wang, Y. et al., China's electric car surge. *Energy Policy,* **102**, pp. 486–490, 2017.
[57] Zhao, X., Zeng, Y. & Zhao, D., Distributed solar photovoltaics in China: Policies and economic performance. *Energy,* **88**, p. 572–583, 2015.
[58] Yuan, X. et al., The development of building energy conservation in China: A review and critical assessment from the perspective of policy and institutional systems. *Sustainability,* **9**(9), pp. 1654, 2017.
[59] Sen, S. et al., Renewable energy scenario in India: Opportunities and challenges. *Journal of African Earth Sciences,* **122**, pp. 25–31, 2016.
[60] Energy Conservation Act, 2001.
[61] The Integrated Energy Policy, 2006.

[62] National Action Plan on Climate Change, 2008.
[63] 2011, Strategic Plan for New and Renewable Energy Sector for the Period 2011 – 2017.
[64] Saha, A. & Talwar, K., India's response to climate change: The 2009 Copenhagen summit and beyond. *National University of Juridical Sciences Law Review,* **3**(2), pp. 159, 2010.
[65] Badrinarayana, D., Transporting India into a new climate: The implications for energy law and policy. *Cumberland Law Review,* **40**(1), pp. 119–133, 2009.
[66] Roy, A.G., Securing India's energy options in an interdependent world. *Renewable Energy Law and Policy,* **3**(2), pp. 131–139, 2012.
[67] Mallah, S. & Bansal, N.K., Renewable energy for sustainable electrical energy system in India. *Energy Policy,* **38**(8), pp. 3933–3942, 2010.
[68] Rosha, A. & Freestone, D., A., Green emerging market: India's experiments with market based mechanisms for climate mitigation. *Carbon and Climate Law Review,* **6**(4), pp. 342–353, 2012.
[69] Electricity Act, 2003.
[70] Tariff Policy, 2006.
[71] Schmid, G., The development of renewable energy power in India: Which policies have been effective? *Energy Policy,* **45**, pp. 317–326, 2012.
[72] Kar, S.K., Renewable energy market developments: A study of India. *Renewable Energy Law and Policy,* **4**, pp. 238, 2015.
[73] Stern, A.C., History of air pollution legislation in the United States. *Journal of the Air Pollution Control Association,* **32**(1), pp. 44–61, 1982.
[74] IEA, Global Renewable Energy, IEA Online. www.iea.org. Accessed on: 16 Dec. 2018.
[75] Air Pollution Control Act, 1955.
[76] EPA, Clean Air Act Text, EPA. www.epa.gov/clean-air-act-overview/clean-air-act-text. Accessed on: 16 Dec. 2018.
[77] Energy Policy Act of, 2005, 119 Stat. 594, Public Law 109-58.
[78] Energy Independence and Securities Act of 2007, 121 Stat. 1492, Public Law 110-140.
[79] American Recovery and Reinvestment Act of 2009, 123 Stat. 115 Public Law 111-5.
[80] Clean Air Act of 2015 (as amended through Public Law 114-94).
[81] American Recovery and Reinvestment Act of 2009, 123 Stat. 115 Public Law 111-5, s. 406.
[82] Energy Independence and Securities Act of 2007, 121 Stat. 1492, Public Law 110-140.
[83] Rogers-Hayden, T., Hatton, F. & Lorenzoni, I., Energy security and climate change: Constructing UK energy discursive realities. *Global Environmental Change,* **21**(1), pp. 134–142, 2011.
[84] Barton, J. et al., Transition pathways for a UK low-carbon electricity system: Comparing scenarios and technology implications. *Renewable and Sustainable Energy Reviews,* **82**, p. 2779–2790, 2018.
[85] Nachmany M. et al., The 2015 global climate change legislation study: A review of climate change legislation in 99 countries. London School of Economics. www.lse.ac.uk. Accessed on: 17 Aug.t 2018.
[86] OECD, IEA & IRENA, United Kingdom Renewable Energy Policy Framework Summary, IEA. www.iea.org. Accessed on: 16 Dec. 2018.

[87] FitzPatrick, D., UK Legislative Framework for Renewable Energy, Taylor Wessing. www.united-kingdom.taylorwessing.com. Accessed on: 20 Dec. 2018.
[88] Climate Change Act, Cap 27, 2008.
[89] Cave, S., *Climate Change Legislation and Targets,* Northern Ireland Assembly Research and Information Service Research Paper 135/13, pp. 4, 2013.
[90] Directive 2009/28/EC of the European Parliament of the Council of 23 April, 2009.
[91] National Renewable Energy Action Plan for the United Kingdom, Article 4 of the Renewable Energy Directive 2009/28/EC at 6 and 114.
[92] DECC, UK Renewable Energy Roadmap Policy Paper, July 2011 at 9.
[93] Energy Act 2013 Cap 32.
[94] Anandarajah, G., Interactions and implications of renewable and climate change policy on UK energy scenarios. *Energy Policy,* **38**(11), pp. 6724–6735, 2010.
[95] McHarg, A., Climate change constitutionalism? Lessons from the United Kingdom. *Climate Law,* **2**, pp. 469–484, 2011.
[96] Hafez, A., Investigating the effectiveness of UK energy policy in promoting renewable investments and reducing carbon emissions. *Zagreb International Review of Economics and Business,* **17**, pp. 1–25, 2014.
[97] Webb, J., Hawkey D. & Tingey, M., Governing cities for sustainable energy: The UK case. *Cities,* **54**, pp. 28–35, 2016.
[98] Liscow, Z. & Karpilow, Q., Innovation snowballing and climate law. *Washington University Law Review,* **95**, p. 387, 2017.
[99] Ranchordas, S., Innovation experimentalism in the age of the sharing economy. *Lewis and Clark Law Review,* **19**(4), pp. 871–924, 2015.
[100] Mormann, F., Enhancing the investor appeal of renewable energy. *Environmental Law,* **42**, p. 681–734, 2012.
[101] Wiseman, H., Expanding regional renewable governance. *Harvard Environmental Law Review,* **35**, pp. 477–614, 2011.

CALCULATION METHOD FOR FIN-AND-TUBE HEAT EXCHANGERS OPERATING WITH NONUNIFORM AIRFLOW

PAOLO BLECICH, ANICA TRP & KRISTIAN LENIĆ
Faculty of Engineering, University of Rijeka, Croatia

ABSTRACT

This paper presents a new calculation method for fin-and-tube heat exchangers operating with nonuniform airflow. The calculation method consists of discretizing the heat exchanger tubes into tube elements and applying mass and energy conservation equations. The calculation method is capable of predicting local and overall values of thermal effectiveness, heat transfer rates and fluid outlet temperatures in fin-and-tube heat exchangers subject to airflow nonuniformity. The results of the tube element method are compared against experimental data, and good accordance has been achieved between predicted and measured values. Experimental data is collected for fin-and-tube heat exchangers placed inside an open wind tunnel, where airflow nonuniformity is generated by partially obstructing the entrance cross section. This study revealed that airflow nonuniformity causes effectiveness deterioration in fin-and-tube heat exchangers. The exact value of effectiveness deterioration depends on the heat exchanger dimensionless groups – the number of heat transfer units (NTU) and the heat capacity rate ratio (C^*) – as well as on the intensity and orientation of the airflow nonuniformity. The design of the tube-side fluid circuitry, which affects the heat exchanger flow arrangement, also plays a role in the effectiveness deterioration. The tube element method revealed that the effectiveness deterioration increases with the intensity of airflow nonuniformity. For an observed nonuniform airflow profile, the effectiveness deterioration is maximum in heat exchangers with a balanced heat capacity rate ratio ($C^* = 1$) and minimum in evaporators and condensers ($C^* = 0$).
Keywords: fin-and-tube heat exchanger, airflow nonuniformity, effectiveness deterioration, calculation method.

1 INTRODUCTION

Fin-and-tube heat exchangers find extensive application in the building sector, in the transport, process and power industries. The inlet airflow distribution strongly affects the performance of fin-and-tube heat exchangers. Nonuniform airflow distribution or airflow nonuniformity causes loss of thermal performance in heat exchangers. Airflow nonuniformity may be induced by poor design of the inlet header, adverse operating conditions, vicinity of the blower, fouling or geometrical irregularities on the heat exchanger surfaces, as stressed by Kitto and Robertson [1], Mueller and Chiou [2], and Shah and Sekulić [3]. Generally, sizing and calculation methods for fin-and-tube heat exchanger assume uniform airflow over the entire heat exchanger surface.

Airflow nonuniformity negatively affects the thermal and hydraulic performance of fin-and-tube heat exchangers, see Singh et al. [4]. The loss of thermal performance is represented by a reduction of the heat flow rate and thermal effectiveness of the heat exchanger. Based on a review of the literature, the general conclusion is that the thermal performance deterioration depends on the number of heat transfer units (NTU), the heat capacity rate ratio (C^*), the tube-side fluid circuitry, the intensity and orientation of the airflow nonuniformity. Furthermore, it was seen that the thermal performance deterioration is higher in heat exchangers with balanced heat capacity rates ($C^* = 1$) than in evaporators and condensers ($C^* = 0$), as referred by Ranganayakulu and Seetharamu [5].

WIT Transactions on Ecology and the Environment, Vol 237, © 2019 WIT Press
www.witpress.com, ISSN 1743-3541 (on-line)
doi:10.2495/ESUS190021

Experimental, analytical and numerical methods have been extensively used to investigate the effect of nonuniform airflow on the performance of fin-and-tube heat exchangers, see Domanski and Yashar [6], Mao et al. [7] and Ranganayakulu and Seetharamu [8]. Generally, nonuniform airflow causes effectiveness deterioration in the range between between 5 and 15%, as stated by Bahman and Groll [9] and Hajabdollahi and Seifoori [10]. The effectiveness deterioration can be even higher, up to 30%, if nonuniform airflow causes refrigerant maldistribution on the tube side, as concluded by Yashar et al. [11]. Generally, the abovementioned research are limited to simple one-dimensional nonuniform airflow profiles or provide data for specific heat exchangers.

This paper presents a new method for the calculation of the thermal performance of fin-and-tube heat exchangers operating in nonuniform airflow, which can be also used for two-dimensional nonuniform airflow profiles. Section 2 shows the analysed heat exchanger, describes the experimental and analytical approaches while Section 3 presents the validation and the results obtained with the presented analytical method.

2 METHODOLOGY

2.1 The fin-and-tube heat exchanger

The analysed fin-and-tube heat exchanger is a three-row heat exchanger with a total 38 tubes, distributed into seven flow circuits (HEX:3R-7C). The seven flow circuits have different number of passes: five flow circuits have a six-pass arrangement and two flow circuits have a four-pass arrangement, as shown in Fig. 1. It can be observed that the fin-and-tube heat exchanger exhibits a Z-shaped flow circuitry on the tube side causing water to flow back-and-forth between the first and third tube row.

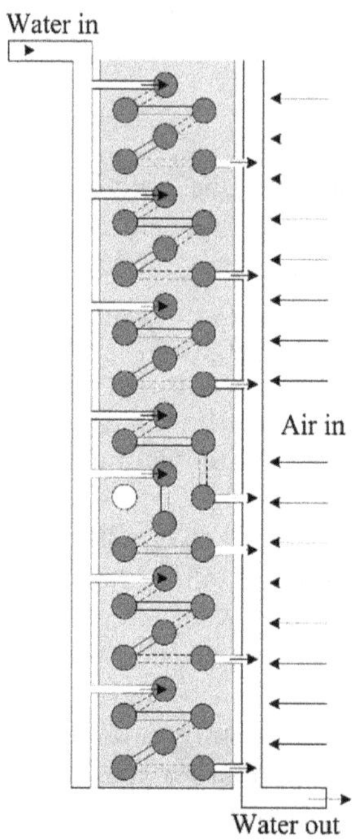

Heat exchanger	HEX:3R-7C
Overall height, mm	840
Overall length, mm	760
Overall depth, mm	155
Tube fluid flow length, mm	700
Fin depth, mm	90
Fin height, mm	780
Number of fins, -	260
Fin pitch, mm	2.6
Fin thickness, mm	0.2
Fin type	flat plain
Fin material	aluminum
Air-side heat transfer area, m^2	40
Surface compactness, m^2/m^3	814
Number of rows	3
Transverse tube pitch, mm	60
Longitudinal tube pitch, mm	30
Outer tube diameter, mm	15.9
Inner tube diameter, mm	14.8
Tube material	copper

Figure 1: Geometry and tube-side flow circuitry of the HEX:3R-7C heat exchanger.

2.2 Experimental method

The experimental setup consists of an open wind tunnel, the heat exchanger test section equipped with temperature and air velocity measurement sensors and a data acquisition system. The experimental setup is shown in Fig. 2.

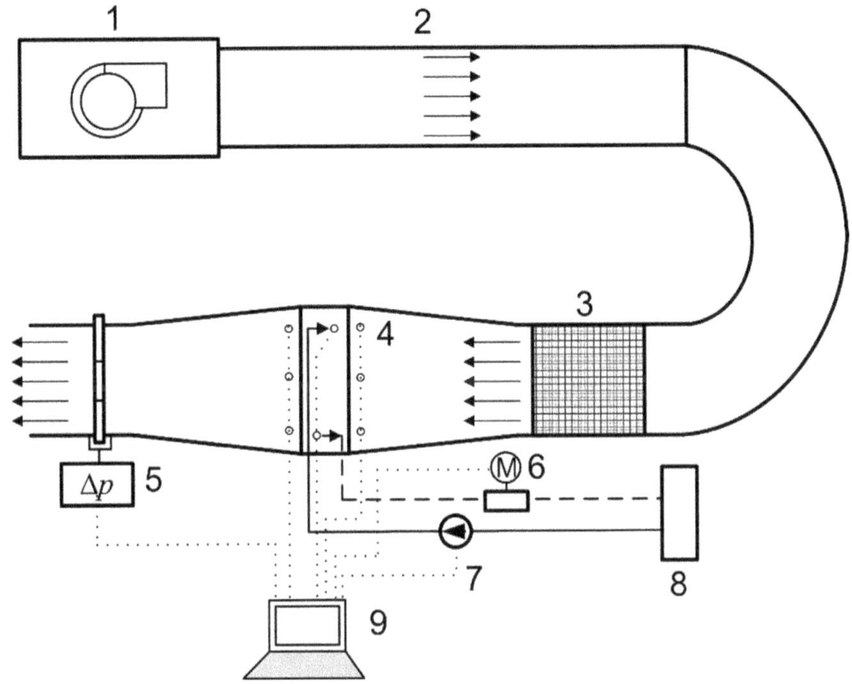

Figure 2: The open wind tunnel. 1: Blower; 2: Air ducts; 3: Flow straightener; 4: Heat exchanger test section; 5: Orifice; 6: Water mass flow meter; 7: Water pump; 8: Heat source; 9: Data acquisition.

The fin-and-tube heat exchanger is placed inside the test section. The air mass flow rate was determined from the pressure drop generated through a calibrated orifice. The water mass flow rate flowing through the heat exchanger pipes was measured using an ultrasound mass flow meter. K-type thermocouples were placed across the heat exchanger inlet and outlet cross section in order to measure the inlet and outlet air temperatures. Inlet and outlet water temperatures were measured using Pt100 temperature sensors.

The placement of the temperature sensors is shown in Fig. 3(a) and (b). Uniform airflow was obtained by leaving the inlet cross section unobstructed, Fig. 3(c), while nonuniform airflow was generated by placing screen and partially obstructing the inlet cross section, as shown in Fig. 3(d)–(f). The water mass flow rate was in the range between 800 and 3200 kg/h, while the air mass flow rate was 1650 kg/h. The air inlet temperature was 15°C while the water inlet temperature was 60°C. The combined uncertainty of all measurements was estimated to be within ±3.0% at 95% confidence level. The average exchanger heat transfer rate ($\dot{Q}_m$) in the fin-and-tube heat exchanger is calculated from the arithmetic mean of the heat transfer rates measured on the air side ($\dot{Q}_a$) and on the water side ($\dot{Q}_w$):

$$\dot{Q}_{\rm m} = \frac{\dot{Q}_{\rm a} + \dot{Q}_{\rm w}}{2}. \quad (1)$$

The heat exchanger effectiveness (ε) is the ratio of the actual heat transfer rate ($\dot{Q}_{\rm m}$) to the maximum theoretical heat transfer rate ($\dot{Q}_{\rm max}$):

$$\varepsilon = \frac{\dot{Q}_{\rm m}}{\dot{Q}_{\rm max}} = \frac{\dot{Q}_{\rm a} + \dot{Q}_{\rm w}}{2\,\dot{m}_{\rm a}\,c_{\rm a}\left(T_{\rm w,in} - T_{\rm a,in}\right)}. \quad (2)$$

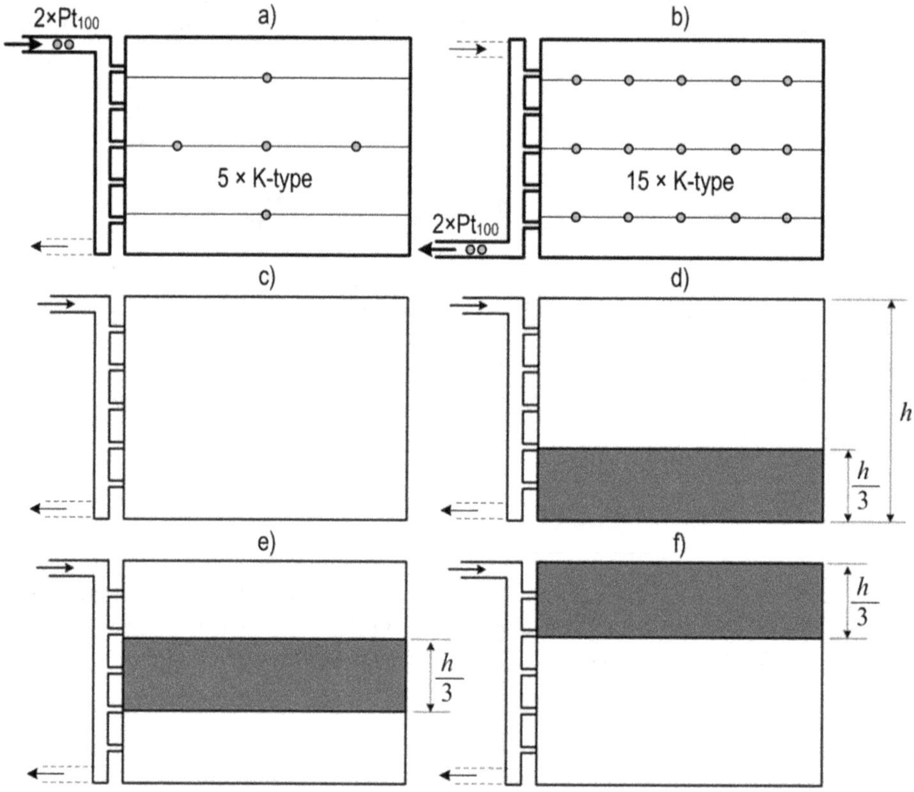

Figure 3: The fin-and-tube heat exchanger setup. (a) Temperature sensors at the air inlet cross section; (b) Temperature sensors at the air outlet cross section; (c) Unobstructed inlet – uniform airflow; (d)–(f) Obstructed inlet with screens – nonuniform airflows, profiles A1, A2 and A3, respectively.

The air and water mass flow rates are $\dot{m}_{\rm a}$ and $\dot{m}_{\rm w}$ while the air and water inlet temperatures are $T_{\rm a,in}$ and $T_{\rm w,in}$, respectively. The air and water specific heat capacities are $c_{\rm a}$ and $c_{\rm w}$. Heat exchangers operating under nonuniform airflow are subject to a loss of thermal performance that can be expressed as effectiveness deterioration. The effectiveness deterioration ($\Delta\varepsilon$) is the relative difference between the heat exchanger effectiveness operating in uniform airflow ($\varepsilon_{\rm uniform}$) and the heat exchanger effectiveness operating in nonuniform airflow ($\varepsilon_{\rm nonuniform}$):

$$\Delta\varepsilon = \frac{\varepsilon_{\text{uniform}} - \varepsilon_{\text{nonuniform}}}{\varepsilon_{\text{uniform}}} \cdot 100\% . \tag{3}$$

2.3 Analytical method

The most widely used thermal design method for fin-and-tube heat exchangers is the effectiveness-number of transfer units method (ε-NTU method). In this method, the heat exchanger is defined by three independent dimensionless groups: (1) the thermal effectiveness (ε); (2) the number of transfer units (NTU); and (3) the heat capacity rate ratio (C^*). Thus, the heat exchanger effectiveness depends on the dimensionless groups NTU and C^* but also on the fluid flow arrangement:

$$\varepsilon = f\left(NTU = \frac{U \cdot A}{C_{\min}},\ C^* = \frac{C_{\min}}{C_{\max}},\ \text{fluid flow arrangement} \right). \tag{4}$$

Mathematical expressions for the heat exchanger effectiveness are available for simple fluid flow arrangements, such as the counterflow, the parallelflow and the crossflow:

$$\text{counterflow:} \begin{cases} \varepsilon = \dfrac{1-\exp[-NTU(1-C^*)]}{1-C^*\exp[-NTU(1-C^*)]}, & \text{for } 0 \le C^* < 1 \\ \varepsilon = \dfrac{NTU}{1+NTU}, & \text{for } C^* = 1 \end{cases}, \tag{5}$$

$$\text{parallelflow:} \quad \varepsilon = \frac{1-\exp[-NTU(1+C^*)]}{1+C^*}, \tag{6}$$

$$\text{crossflow:} \quad \varepsilon = 1 - \exp\left\{ \frac{NTU^{0.22}\left[\exp(-NTU^{0.78}C^*) - 1\right]}{C^*} \right\}. \tag{7}$$

It should be noted that the effectiveness of the crossflow heat exchanger is expressed by the Eckert approximation (eqn (7)), which deviates from the exact Nusselt solution by ±3%. For single-pass crossflow fin-and-tube heat exchangers with less than 4 tube rows the Eckert formula overestimates the heat exchanger effectiveness. For a single-pass three tube row heat exchanger with $C_{\min}$ on the air-side [12] the following expression applies:

$$\left.\begin{array}{r}\text{single-pass three}\\ \text{tube row crossflow}\end{array}\right\} \begin{array}{c} \varepsilon = \left\{1-\exp(-3KC^*)\left[1+C^*K^2(3-K)+3C^{*2}K^4/2\right]\right\}/C^* \\ K = 1-\exp(-NTU/3) \end{array}. \tag{8}$$

The analysis becomes increasingly difficult for complex fluid flow arrangement and analytical solutions are available for a limited number of arrangements, as provided by Bačlić and Gvozdenac [13], but often include infinite series and complex mathematical functions. Boundary conditions, specifically the mass flow rates and inlet temperatures of the two fluids, are necessary to achieve closure in ε-NTU equations.

The heat capacity rate ratio (C^*) is easily determined from the mass flow rates and specific heat capacities of the two fluids, but calculation of the number of transfer units ($NTU = UA/C_{\min}$) is more complex since it includes the overall thermal conductance (UA), which is the product of the overall heat transfer coefficient (U) and the total heat transfer area (A):

$$UA = \left[\frac{1}{(\eta hA)_a} + \frac{\ln(d_o/d_i)}{2\pi(kLN)_t} + \frac{1}{(hA)_w} \right]^{-1}. \tag{9}$$

The water heat transfer coefficient (h_w) is calculated using the correlation of Gnielinski [14], and the air-side heat transfer coefficient (h_a) is determined from the Wang et al. [15] correlation for the Colburn j factor. The surface efficiency on the air-side (η_a) is determined from the Schmidt approximation method [16].

A fin-and-tube heat exchanger with complex fluid flow arrangement and nonuniform fluid temperatures and velocities at the inlet can be analysed by the tube element method (TEM). In the tube element method, the fin-and-tube heat exchanger is discretized into tube elements, on which mass and energy conservation equations can be applied. At the entrance of each tube element, the fluid flow velocities and temperatures are assumed uniform. By performing a fine discretization it can be assumed that the heat capacity rate of the tube-side is much bigger than the heat capacity rate of the air-side ($C_{min,el} / C_{max,el} \approx 0$). This is a fair assumption, especially when the tube side fluid is water or a refrigerant. In that case, the thermal effectiveness of the element can be determined as the thermal effectiveness of an ideal evaporator or condenser.

Another assumption is that the tube-side fluid temperature changes linearly along the tube element. In that case the average temperature of the tube-side fluid is the arithmetic mean of the inlet and outlet temperatures at the tube element. Fig. 4 shows the idea behind the tube element method. For the observed tube element, the air mass flow rate is denoted with $\dot{m}_{a,el}$, the inlet and outlet air temperature with $T_{a,el,in}$ and $T_{a,el,out}$. On the tube side, the water mass flow rate is denoted with $\dot{m}_{w,el}$, the inlet and outlet water temperatures with $T_{w,el,in}$ and $T_{w,el,out}$.

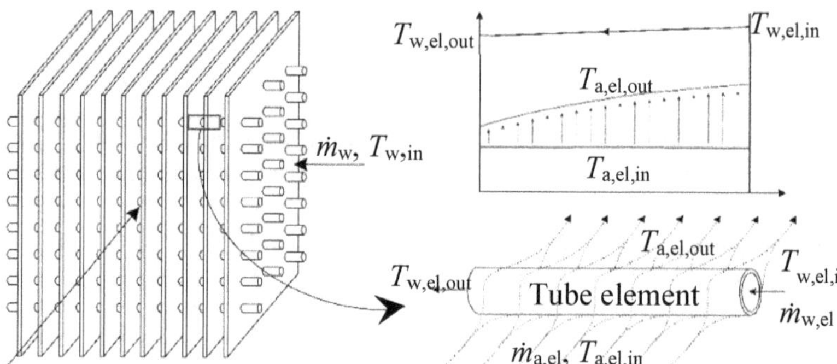

Figure 4: The fin-and-tube heat exchanger and its discretization into tube elements.

The heat balance equation for an observed tube element is written as:

$$\dot{m}_{a,el}\, c_a \left(T_{a,el,out} - T_{a,el,in}\right) = \dot{m}_{w,el}\, c_w \left(T_{w,el,in} - T_{w,el,out}\right). \tag{10}$$

The air and water mass flow rates at the tube element are $\dot{m}_{a,el}$ and $\dot{m}_{w,el}$. The air and water inlet and outlet temperatures are $T_{a,el,in}$, $T_{a,el,out}$, $T_{w,el,in}$, $T_{w,el,out}$, respectively. The heat capacity rate ratio for the tube element (C_{el}*) is calculated from:

$$C_{\mathrm{el}}^{*}=\frac{\dot{m}_{\mathrm{a,el}}\,c_{\mathrm{a}}}{\dot{m}_{\mathrm{w,el}}\,c_{\mathrm{w}}}\ . \tag{11}$$

Under nonuniform airflow conditions, each of the tube elements is exposed to different air mass flow rates. In that case, the air mass flow rate at the tube elements can be expressed using the continuity equation:

$$C_{\mathrm{el}}^{*}=\frac{w_{\mathrm{a,el}}A_{\mathrm{a}}\rho_{\mathrm{a}}c_{\mathrm{a}}}{\dot{m}_{\mathrm{w,el}}c_{\mathrm{w}}}\ . \tag{12}$$

In the above equation, the air velocity at the tube element is $w_{\mathrm{a,el}}$, the tube element cross section is A_{a} and the air density ρ_{a}. On the tube-side the water mass flow rate at the tube element is determined by dividing the total water mass flow rate by the number of tube circuits, assuming that the water flow is uniformly distributed among them. The number of transfer units of the tube element (NTU_{el}) is determined from:

$$NTU_{\mathrm{el}}=\frac{UA_{\mathrm{el}}}{C_{\mathrm{a}}}=\frac{UA_{\mathrm{el}}}{w_{\mathrm{a,el}}A_{\mathrm{a}}\rho_{\mathrm{a}}c_{\mathrm{a}}}\ . \tag{13}$$

The tube element effectiveness ($\varepsilon_{\mathrm{el}}$) is calculated assuming that $C^* = 0$, that is:

$$\varepsilon_{\mathrm{el}}=1-\exp(-NTU_{\mathrm{el}})\ . \tag{14}$$

The air and water temperatures at the outlet of the tube element can be determined from the system of equations comprising the tube element heat balance (eqn (10)) and the tube element dimensionless groups (eqns (12)–(14)):

$$\begin{aligned} T_{\mathrm{w,el,out}}&=\frac{(2-\varepsilon_{\mathrm{el}}C_{\mathrm{el}}^{*})T_{\mathrm{w,el,in}}+2\varepsilon_{\mathrm{el}}C_{\mathrm{el}}^{*}T_{\mathrm{a,el,in}}}{2+\varepsilon_{\mathrm{el}}C_{\mathrm{el}}^{*}} \\ T_{\mathrm{a,el,out}}&=\frac{2\varepsilon_{\mathrm{el}}T_{\mathrm{w,el,in}}+\left[\varepsilon_{\mathrm{el}}(C_{\mathrm{el}}^{*}-2)+2\right]T_{\mathrm{a,el,in}}}{2+\varepsilon_{\mathrm{el}}C_{\mathrm{el}}^{*}} \end{aligned}\ . \tag{15}$$

The TEM method can calculate the heat exchanger effectiveness by taking into account the effects of nonuniform airflow. Furthermore, it can take into account the effects of complex tube-side flow circuitry, beyond the reach of available analytical eqns (5)–(8). The number of tube elements used in the TEM method is selected depending on the desired accuracy and the available processor resources. A numerical algorithm for the TEM method was developed. The necessary input parameters are the geometry of the tube-side circuitry, the tube connecting sequence and the number of tubes in each circuit as well as the airflow profile at the inlet. Each tube element in the heat exchanger is assigned three indexes: the first one for the tube element sequence number in the observed tube, the second for the tube sequence number in the observed flow circuit and the third for the flow circuit sequence number in the heat exchanger. The TEM algorithm follows the direction of the tube-side fluid flow, with the outlet temperature of the current tube element being the inlet temperature for the following tube element. Overall, the heat exchanger can be discretized into thousands of tube elements onto which mass and energy conservation equations are applied and solved until reaching the desired level of solution accuracy.

3 RESULTS AND DISCUSSION

3.1 Validation of the analytical method

The results of the TEM method are validated against measurement data collected on the HEX:3R-7C heat exchanger under nonuniform airflow (Fig. 5). The heat exchanger operating conditions are defined by air and water mass flow rates of $\dot{m}_a = 1650$ kg/h and $\dot{m}_w = 800 - 3200$ kg/h, with air and water inlet temperatures of $t_{a,in} = 15$°C and $t_{w,in} = 60$°C, respectively. Overall, the TEM method achieves very good agreement with experiments. The TEM underestimates the heat exchanger effectiveness in the region of $C^* > 0.4$. Tube-side flow nonuniformity and fin longitudinal heat conduction, both neglected by the TEM method, could be among the possible causes of this discrepancy.

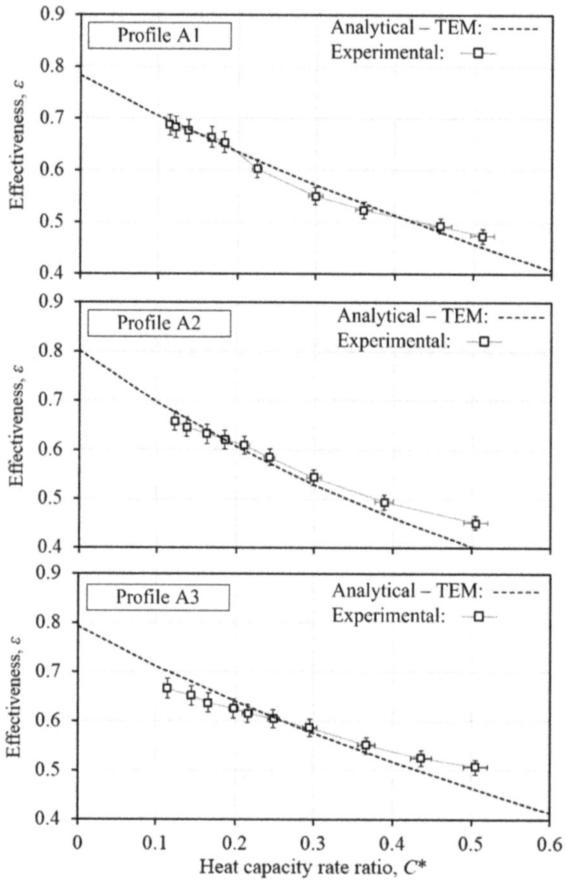

Figure 5: TEM validation: HEX:3R-7C fin-and-tube heat exchanger in nonuniform airflow profiles A1, A2 and A3.

The TEM method is first used to calculate the effectiveness of the HEX:3R-7C heat exchanger with uniform airflow conditions. This is done in order to understand the effect of the flow arrangement on the heat exchanger effectiveness. Fig. 6 shows the heat exchanger effectiveness calculated with the TEM method and compared against the thermal effectiveness of the pure parallelflow heat exchanger (eqn (6)). Fig. 7 shows the comparison between the effectiveness of the HEX:3R-7C heat exchanger and the effectiveness of the single-pass three-tube-row crossflow heat exchanger (eqn (8)). It can be seen that the effectiveness of the HEX:3R-7C heat exchanger is much closer to the effectiveness of the pure parallelflow (Fig. 6) heat exchanger than to the effectiveness of the single-pass three-tube-row crossflow heat exchanger (Fig. 7). This suggests that the Z-shaped tube circuitry of the HEX:3R-7C heat exchanger induces a parallel-crossflow arrangement between the fluid and a consequent loss of thermal performance.

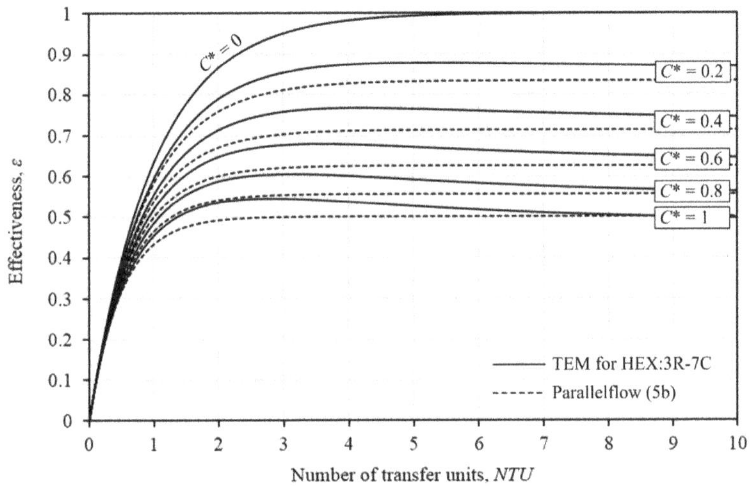

Figure 6: TEM results for the HEX:3R-7C heat exchanger under uniform airflow – comparison against the parallelflow.

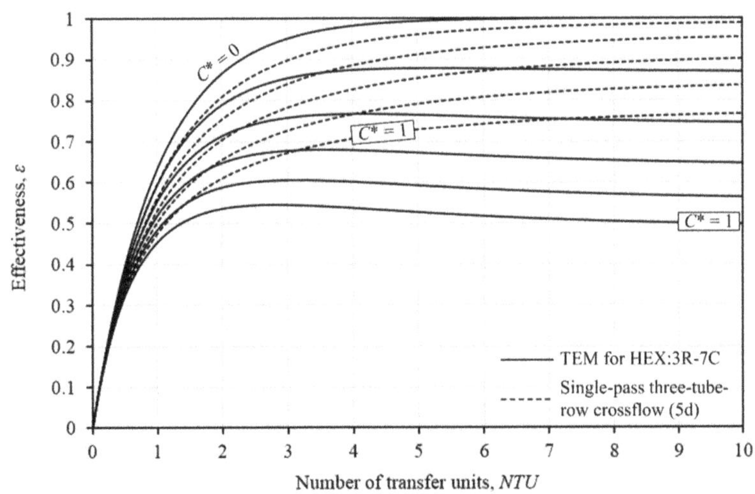

Figure 7: TEM results for the HEX:3R-7C heat exchanger under uniform airflow – comparison against the single-pass three-row crossflow (eqn (8)).

3.2 Heat exchanger effectiveness for nonuniform airflow

The effectiveness of the HEX:3R-7C heat exchanger under nonuniform airflow conditions is calculated using the TEM method. Fig. 8 shows the effectiveness of the HEX:3R-7C heat exchanger operating under nonuniform airflow profile A1 and the comparison against the effectiveness obtained for uniform airflow. Nonuniform airflow profile A1 was generated by blocking access to airflow to the two lowermost flow circuits. Fig. 9 plots the effectiveness deterioration caused by nonuniform airflow A1. The effectiveness deterioration is calculated as the relative difference between the heat exchanger effectiveness for uniform and nonuniform airflow profile, as stated by eqn (3).

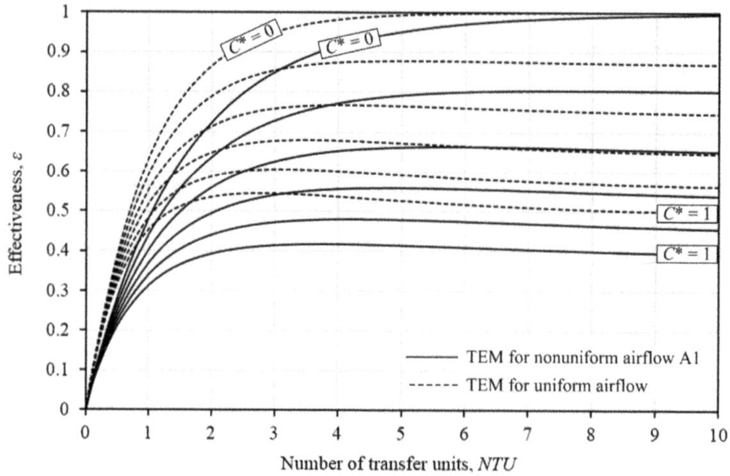

Figure 8: TEM results for the HEX:3R-7C heat exchanger under nonuniform airflow profile A1 – comparison against TEM results for uniform airflow.

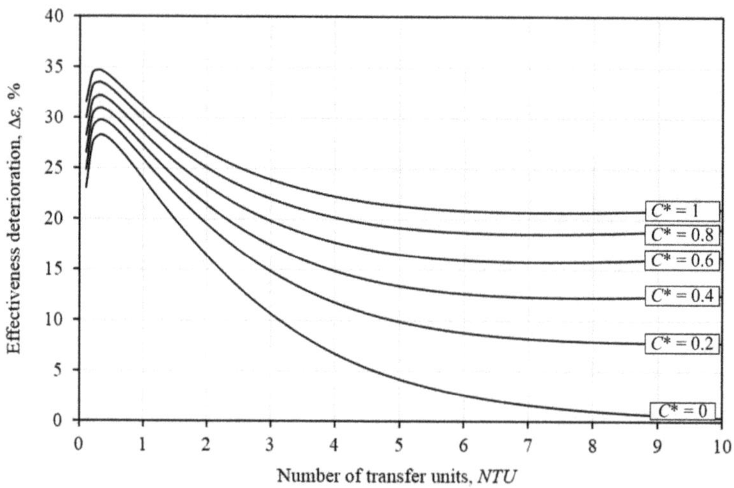

Figure 9: TEM results: effectiveness deterioration caused by nonuniform airflow profile A1.

4 CONCLUSION

A new calculation method for fin-and-tube heat exchangers was presented in this paper – the TEM method. The TEM method was validated against experiments carried out on a heat exchanger operating under nonuniform airflow and very good agreement between predicted and measured values of heat exchanger effectiveness was achieved. The TEM method can be seen as an extension of the standard ε-NTU thermal design method but with the addition of being able to determine the effectiveness of heat exchangers in nonuniform airflow and with complex tube-side flow circuitry. The test subject was a three-row, seven flow circuits fin-and-tube heat exchanger with a total of 38 tubes. The tube-side flow circuitry exhibits a Z-shaped pattern, with water flowing back-and-forth with respect to the incoming airflow. The TEM method revealed that the Z-shaped flow circuitry generates a loss of thermal performance and the resulting effectiveness curves resemble closely the curves of the pure parallelflow rather than those of the single-pass three-row crossflow. Further, the TEM method revealed that nonuniform airflow causes effectiveness deterioration. For the observed nonuniform airflow profiles, the size of the effectiveness deterioration depended on the heat exchanger dimensionless groups, the heat capacity rate ratio C^* and the number of transfer units NTU. The method showed that the effectiveness deterioration is the biggest for balanced heat capacity rate ratios ($C^* = 1$), decreases with the decrease of C^* and is the smallest in evaporators and condensers ($C^* = 0$).

The TEM method proved as a valuable tool for the thermal design of fin-and-tube heat exchangers capable of improving the circuitry of the tube-side fluid flow and assessing the effects of airflow nonuniformity on the thermal performance of the heat exchanger.

ACKNOWLEDGEMENT

This work has been fully supported by the Croatian Science Foundation under the project IP-2016-06-4095.

REFERENCES

[1] Kitto, J.B. & Robertson, J.M., Effects of maldistribution of flow on heat transfer equipment performance. *Heat Transfer Engineering,* **10**(1), pp. 18–25, 1989.

[2] Mueller, A.C. & Chiou, J.P., Review of various types of flow maldistribution in heat exchangers. *Heat Transfer Engineering,* **9**(2), pp. 36–50, 1988.

[3] Shah, R.K. & Sekulić, D.P., *Fundamentals of Heat Exchanger Design*, John Wiley & Sons, Inc.: Hoboken, 2003.

[4] Singh, S.K., Mishra, M. & Jha, P.K., Nonuniformities in compact heat exchangers – scope for better energy utilization: A review. *Renewable and Sustainable Energy Reviews*, **40**, pp. 583–596, 2014.

[5] Ranganayakulu, C. & Seetharamu, K.N., The combined effects of wall longitudinal heat conduction, inlet fluid flow nonuniformity and temperature nonuniformity in compact tube-fin heat exchangers: a finite element method. *International Journal of Heat and Mass Transfer*, **42**(2), pp. 263–273, 1999.

[6] Domanski, P.A. & Yashar, D.A., Optimization of finned-tube condensers using an intelligent system. *International Journal of Refrigeration*, **30**(3), pp. 482–488, 2007.

[7] Mao, J.N., Chen, H.X., Jia, H., Wang, Y.Z. & Hu, H.M., Effect of air-side flow maldistribution on thermal-hydraulic performance of the multi-louvered fin and tube heat exchanger. *International Journal of Thermal Sciences*, **73**, pp. 46–57, 2013.

[8] Ranganayakulu, C. & Seetharamu, K.N., *Compact Heat Exchangers: Analysis, Design and Optimization using FEM and CFD Approach*, John Wiley & Sons: USA, 2018.

[9] Bahman, A.M. & Groll, E.A., Application of interleaved circuitry to improve evaporator effectiveness and COP of a packaged AC system. *International Journal of Refrigeration*, **79**, pp. 114–129, 2017.
[10] Hajabdollahi, H. & Seifoori, S., Effect of flow maldistribution on the optimal design of a cross flow heat exchanger. *International Journal of Thermal Sciences*, **109**, pp. 242–252, 2016.
[11] Yashar, D.A., Domanski, P.A. & Cho, H.H., An experimental and computational study of approach air distribution for a finned-tube heat exchanger. *HVAC&R Research*, **17**(1), pp. 76–85, 2011.
[12] ESDU, Effectiveness-NTU Relationships for the Design and Performance Evaluation of Two-Stream Heat Exchangers, 167 86018, Amendment A, Engineering Science Data Unit Int.: London, pp. 92–107, 1991.
[13] Bačlić, B.S. & Gvozdenac, D.D., Exact explicit equations for some two- and three-pass cross-flow heat exchangers effectiveness. *Heat Exchangers – Thermal Hydraulic Fundamentals and Design*, eds S. Kakac, A.E. Bergles & F. Mayinger, Hemisphere Publishing Corporation: New York, pp. 481–494, 1981.
[14] Gnielinski, G., New equation for heat and mass transfer in turbulent pipe and channel flow. *International Journal of Chemical Engineering*, **16**, pp. 359–368, 1976.
[15] Wang, C.-C., Chi, K.-Y. & Chang, C.-J., Heat transfer and friction characteristics of plain fin-and-tube heat exchangers, part II: Correlation. *International Journal of Heat and Mass Transfer*, **43**(15), pp. 2693–2700, 2000.
[16] Schmidt, T.E., Heat transfer calculations for extended surfaces. *Refrigerating Engineering*, **57**, pp. 351–357, 1949.

STUDY ON CONTAINER SHIP ENERGY CONSUMPTION

ABDALLAH AIJJOU, LHOUSSAIN BAHATTI & ABDELHADI RAIHANI
Laboratory: Signals, Distributed Systems and Artificial Intelligence (SSDIA) ENSET Mohammedia, University Hassan II, Morocco

ABSTRACT

The shipping industry is facing three challenges: climate change, increasing bunker fuel price and tightening international rules on pollution and CO_2 emissions. All three challenges can be met by reducing fuel consumption and improving global ship energy efficiency. This paper aims to analyze medium size container vessel energy based on the data collected from the sample ship during two regular voyages. The analysis covers the exergy and energy balance of the main components. Container vessels consume the most fuel of the largest fuel oil consumers as they have the most powerful engines. The propulsion is responsible for 82% of the energy demand on a container vessel, electric power production accounts for 17%, while steam generation is limited to 1%. It has been demonstrated that up to 64% of the ship energy consumption is lost to the environment by the cooling system and exhaust gas from propulsion engine and diesel generators. Exergy losses due to the exhaust gas and heat transfer are other contributors. By focusing studies on these items, we may be able to determine how the ship energy efficiency may be improved.
Keywords: container vessel, energy efficiency, energy, exergy.

1 INTRODUCTION

Maritime transport emits around 1,000 million tons of CO_2 annually and is responsible for about 2.5% of global greenhouse gas emissions [1]. Shipping emissions are predicted to increase between 50% and 250% by 2050, depending on future economic and energy developments. The International Maritime Organization (IMO) is in the development of two indices to create a standard to monitor the emissions and stimulate the improvement with interventions both in the design and in the operational phases. The Energy Efficiency Design Index (EEDI, IMO 2012) measures the CO_2 emissions in relation to a nominal transportation work rate and is related to the ship technical characteristics. The Energy Efficiency Operational Index (EEOI, IMO 2012) strictly depends on the cargo carried in a specific voyage and its length [2]. The IMO resolution MEPC.278 (70) amended MARPOL Annex VI on 1 March 2018 to introduce the new Regulation 22A, which includes a requirement for ships to record and report fuel oil consumption data. Ships over 5000 gross tons trading from 1 January 2018 at EU maritime ports are to monitor and later report their related CO_2 emissions in accordance with their monitoring plans [3].

Measures for the improvement of ship energy efficiency are normally divided into design and operational measures. While the former has been associated to larger saving potential, the latter can still provide a significant reduction in fuel consumption, while requiring a much more limited capital investment. However, the large number of variables influencing ship energy efficiency makes it hard to assess ship performance in relation to a standard baseline. Operational measures include, among others, improvement in voyage execution, reduction of auxiliary power consumption, weather routing, optimized hull and propeller polishing schedule, slow steaming, and trim optimization; many literatures have addressed this issue [4]–[8]. Some other studies focused on the energy performances of the individual machinery component [9], [10]. The main fuel oil consumers onboard container vessels are propulsion engines and auxiliary engines, which are the internal combustion diesel type; the auxiliary boiler fuel consumption may be neglected compared to propulsion engine.

www.witpress.com, ISSN 1743-3541 (on-line)
doi:10.2495/ESUS190031

The aim of this paper is to assess the ship energy system performance by analyzing the data provided by ship's records and to evaluate the actual fuel consumption and potential energy saving for better ship energy efficiency. We take as a sample a container vessel of medium size category which represents 20% of the total world container fleet and 30% of the power installed and fuel consumed by the container vessels [11]. The daily report of the ship (noon report) transmits several parameters to the teams ashore; namely, the fuel consumption, the speed of the vessel, the average draught, the loading, the weight of the cargo, the speed and the wind sector, the engine power, electric power, etc. These data are sent every day regardless of the vessel's position (at sea or at the port). Sorting these data was necessary, without, our calculations could be altered because of the zero speed or consumption values corresponding to periods of stop or anchorage. We filter the data and ignore the non-consistent values recorded by mistake or inaccuracy of the crew.

Even though exergy analysis is similar to energy analysis in characteristics, there are some fundamental differences between them [12]. While the energy analysis is based on the first law of thermodynamics (energy conservation), exergy analysis is based on the second law (degradation of energy). Degradation of energy means the loss of exergy systems from the irreversibility. Using only energy analysis is not adequate to analyze energy utilization processes and, additionally, it leads to poor decisions about them. Exergy analysis, more complicated than energy analysis, determines the portion of energy which cannot be used, and it states where the losses occur in a system. By applying it to a thermal system in addition to the energy analysis, irreversibility could be detected and so the best ways to minimize energy destruction and losses and increase efficiency can be understood. Energy analysis will provide the amount of energy variation as a function of energy transfer across the boundaries as heat or work which is related to mass flow passing these boundaries.

2 ENERGY AND EXERGY THEORY

2.1 Diesel engine

Internal combustion engines [13] are mainly used in shipbuilding for propulsion and power generation purposes. As depicted in Fig. 1, the diesel engine converts the chemical energy to mechanical energy in the form of shaft rotation by burning fuel oil; part of the energy is transferred to the atmosphere by exhaust gas and a cooling system.

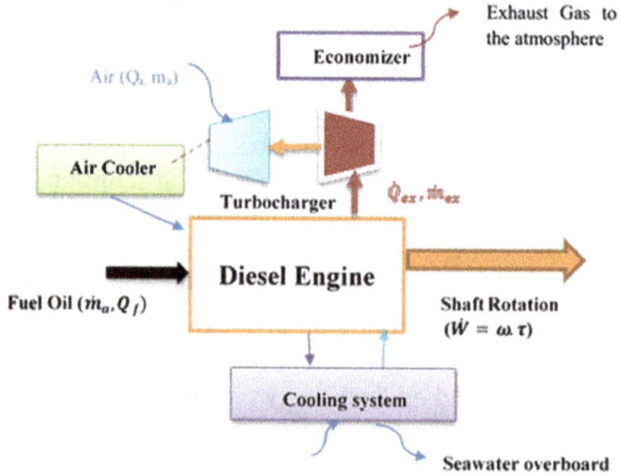

Figure 1: Diesel engine flows.

2.1.1 Energy flow

Energy analysis provides a calculation of the internal energy variation as a function of energy transfers across the boundaries as heat or work and the enthalpy which is related to the mass flow passing these boundaries [14]. Before applying it to the test engine, the following assumptions must be made to simplify the calculations:

- The engine is in a steady-state condition.
- It is assumed that the system is an open system and the reference state is defined as $T_0 = 293$ K and $P_0 = 1$ atm.
- The combustion air and exhaust gases are ideal gas mixtures.
- The potential and kinetic energy effects of the incoming fluid streams and outgoing fluid streams are ignored [15].

In general, work and heat can be expressed as eqn (1):

$$W = \dot{m}(\mathrm{Ke} + \mathrm{Pe}) + W_d\,. \quad Q = \dot{m}\mathrm{h}\,.\,Q_d, \tag{1}$$

where $\dot{m}$ is the mass flow rate carrying the kinetic K_e and potential P_e energy work or enthalpy h.

W_d, Q_d is the direct input/output energy to/from the system.

Mass balance for the control volume in steady-state condition can be written as eqn (2):

$$\sum \dot{m}_i \quad = \sum \dot{m}_o, \tag{2}$$

where $\dot{m}_i$ represents the inlet flow rate of the mass that consists of air and fuel and $\dot{m}_o$ represents outlet mass consisting of exhaust gases. Considering that fuel enters the engine, with mass flow $\dot{m}_f$ rate and is mixed with air $\dot{m}_a$ to form $\dot{m}_i$.

Energy balance for the control volume in steady-state conditions, kinetic and potential energy being neglected, is given by eqn (3) in the general sense:

$$\dot{Q} + \dot{W} = \sum \dot{m}_o h_o - \dot{\sum} \dot{m}_\iota h_\iota, \tag{3}$$

where $\dot{m}$ is the mass flow rate; h is the specific enthalpy. Besides, Q indicates the net heat transfer rate and W indicates the brake power. Considering that the diesel engine generates brake power and some heat produced by combustion is transferred to the environment, eqn (3) can be written as eqn (4), where Q_f is the fuel energy rate, $\dot{Q}_{ex}$ is the exhaust energy rate, $\dot{Q}_d$ is the lost energy rate which is transferred to the environment by cooling, lubricating, heat transfer except for the exhaust:

$$\dot{Q}_f = \dot{W} + \dot{Q}_{ex} + \dot{Q}_d. \tag{4}$$

The brake power is denoted by eqn (5):

$$\dot{W} = \omega.\tau, \tag{5}$$

where (w) is the angular velocity; and (τ) is the engine torque.

Since the combustion air is in the same state as the standard reference state defined as $T_0 = 293$ K and $P_0 = 1$ atm, the energy of the combustion air can be neglected so the heat input rate to the control volume by mass is represented only by chemical energy of fuel $\dot{E}_f$ which can be calculated using the mass flow rate of the fuel $\dot{m}_f$ and Net calorific value of fuel as expressed by eqn (6):

$$\dot{E}_f = \dot{m}_f.NCV. \tag{6}$$

Exhaust energy rate ($\dot{Q}_{ex}$) can be calculated as a function of the mass flow rate of each exhaust gas components ($\dot{m}_{ex}$) which is obtained using combustion equations and the enthalpy change (Δh) represented by the difference between the enthalpy of the exhaust temperature and the reference temperature of each exhaust gas species as given in eqn (7):

$$\dot{Q}_{ex} = \dot{m}_{ex}.\Delta h. \quad (7)$$

The engine characteristics are essentially expressed by thermal efficiency (γ) and brake specific fuel consumption (Sfoc). While thermal efficiency is determined as the ratio of brake power to the fuel energy rate of the control volume, "Sfoc" is defined as a measure of how much fuel is consumed in one hour to obtain one-kilowatt brake power and they are denoted as in eqns (8) and (9):

$$\gamma = \frac{\dot{W}}{\dot{E}_f}, \quad (8)$$

$$\text{Sfoc} = \frac{\dot{m}_f}{\dot{W}}. \quad (9)$$

2.1.2 Exergy flow

The assumptions made for energy analysis is also applicable to exergy analysis. Based on these assumptions, the exergy balance for the control volume can be given in the general sense by eqn (10):

$$\dot{B}_Q + \dot{B}_W - \dot{B}_d = \sum \dot{m}_o \psi_o - \sum \dot{m}_\iota \psi_\iota, \quad (10)$$

where $\dot{B}_Q$ is the exergy transfer rate related to heat transfer between the control volume and the environment; $\dot{B}_w$ is the exergy transfer rate associated with work transfer; $\dot{B}_d$ is the exergy destruction rate of the control volume, and ($\dot{m}_o \dot{\psi}_o / \dot{m}_\iota \psi_i$) are the exergy transfer through the intake and exhaust process, where (ψ_i / ψ_o) are the specific exergies of the fuel and exhaust gas.

Similarly ($\dot{m}_i$ $\dot{m}_o$) are the mass flow rates of the fuel and exhaust gas. Exergy input rate with mass transfer to the engine is formed mainly by fuel exergy, the combustion air exergy can be neglected by assuming that intake air enters the engine at ambient conditions. The specific chemical exergy of fuel is obtained by multiplying the net calorific value of fuel and chemical exergy factor as per eqn (11):

$$\psi_f = \varphi\, NCV. \quad (11)$$

The chemical exergy factor φ is based on the composition of fuel oil (C, H, O, S) [16]. The exergy transfer rate associated with work is equal to the network for the engine, eqn (12):

$$\dot{B}_w = \dot{W}. \quad (12)$$

Exergy output rate with mass transfer from the engine consists of exhaust exergy which can be expressed as in eqn (13):

$$\dot{B}_{ex} = \dot{m}_{ex}\psi = \dot{m}_{ex}\, C_{pex}\, T_0 \left(\frac{T}{T_0} - 1 - In\frac{T}{T_0}\right). \quad (13)$$

Exhaust exergy per unit mass (ψ) can be expressed as the sum of the specific thermomechanical and chemical exergies of the exhaust gases with the assumption that the exhaust gas is an ideal gas mixture [17].

Exergy transfer rate associated with the heat transfer is obtained by eqn (14):

$$\dot{B}_Q = \sum(1 - \frac{T_0}{T})\dot{B}_d. \quad (14)$$

T is the cooling water temperature equal to the system boundary temperature.

$\dot{B}_d$ is the lost (wasted) energy rate which is transferred to the environment by cooling, lubricating, heat transfer except for the exhaust gas.

The exergy efficiency is expressed as the ratio of the brake power exergy to the fuel exergy of the control volume, eqn (15):

$$\gamma = \frac{\dot{W}}{\dot{B}_f}. \quad (15)$$

2.2 Boiler

One oil-fired boiler is generally installed onboard container vessels for heating systems of fuel oil, lubricating oil, cooling water, sanitary and air conditioning. The boiler energy flows are represented in Fig. 2.

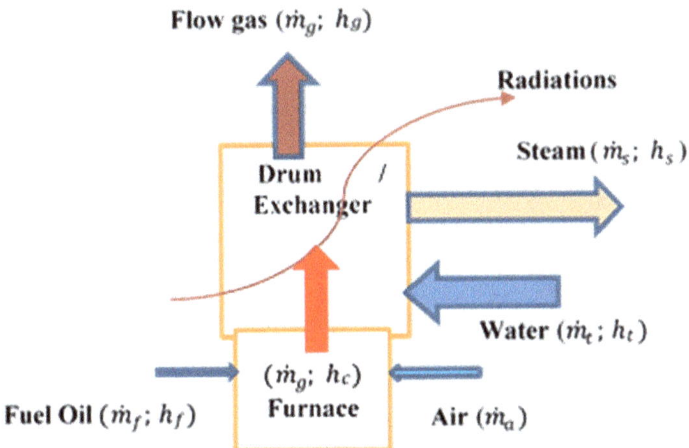

Figure 2: Diagram boiler energy flow. Specific enthalpy (h); specific exergy (ψ); mass flow ($\dot{m}$). Subscripts: air (a); combustion (c); fuel oil (f); flow gas (g); incoming (i); outgoing (o); steam (s); water (t).

The boiler is usually well insulated so no heat dissipation to the surroundings, nor any kind of work is generated. Also, the kinetic and potential energies of the fluid's streams are usually negligible. Then only total energies of the incoming streams and the outgoing mixture will be considered for this study.

2.2.1 Energy analysis

In the boiler furnace, we assume that the combustion is complete and take place at constant pressure, and therefore the total chemical energy of fuel oil is transferred to the flue gases.

Assumptions: Steady operating conditions exist. Kinetic and potential energy changes are negligible. Air properties are used for exhaust gases; pressure drops in the heat exchanger are negligible.

www.witpress.com, ISSN 1743-3541 (on-line)

The mass flow of flue gases is the sum of the fuel oil (f) mass follow and combustion air (a) mass flow (eqn (16)):

$$\dot{m}_g = \dot{m}_f + \dot{m}_a. \tag{16}$$

We assume that the mass flow of the steam produced is equal to the mass flow of the feeding water $\dot{m}_s = \dot{m}_t$.

The heat from combustion gas is transmitted to the water to produce steam. The energy flow $\dot{Q}$ is given by eqn (17):

$$\dot{Q} = \dot{m}_g\left(h_c - h_g\right) + \dot{m}_s(h_s - h_t). \tag{17}$$

2.2.2 Exergy analysis

The maximum paid work of the boiler is defined from exergy balance considering an environment of: T_0 =293 K and P_0 = 1 atm.

As an open system, the exergy balance can be expressed as per eqn (18):

$$\dot{B}_i - \dot{B}_o = \dot{B}_w. \tag{18}$$

$\dot{B}_i$ and $\dot{B}_o$ represent the exergy flow entering and leaving the boiler.

$\dot{B}_w$ denotes the total exergy destruction in the boiler, eqn (21).

Combustion process: ($\dot{B}_{wc}$)is the exergy destroyed (wasted) during fuel oil combustion, expressed in accordance with eqn (19):

$$\dot{B}_{wc} = \dot{m}_f.\psi_f + \dot{m}_a.\psi_a - \dot{m}_g.\psi_g. \tag{19}$$

Heat exchange: ($\dot{B}_{wh}$)is the exergy destroyed (wasted) during heat exchange gas/water, expressed in accordance with eqn (20):

$$\dot{B}_{wh} = \dot{m}_g\left(\psi_c - \psi_g\right) + \dot{m}_s(\psi_t - \psi_s), \tag{20}$$

$$\dot{B}_w = \dot{B}_{wc} + \dot{B}_{wh}, \tag{21}$$

$$\psi_i = h_i - \dot{T}_0\, s_i, \tag{22}$$

where h: enthalpy. s: entropy. T_0: 298°K.

In Table 1, we summarize the different efficiency formulas for the boiler energy and exergy flows.

Table 1: Summary of boiler energy flow efficiencies.

	Energy	Exergy
Combustion	$\eta_c = \dfrac{\dot{m}_g h_c}{\dot{m}_f h_f}$	$\mu_c = \dfrac{\dot{m}_g.\psi_c}{\dot{m}_f.\psi_f}$
Heat exchange	$\eta_h = \dfrac{\dot{m}_s(h_s - h_t)}{\dot{m}_f h_f}$	$\mu_h = \dfrac{\dot{m}_s(\psi_s - \psi_t)}{\dot{m}_g(\psi_c - \psi_g)}$
Boiler	$\eta = \dfrac{\dot{m}_s(h_c - h_t)}{\dot{m}_f h_f}$	$\mu = \dfrac{\dot{m}_s(\psi_s - \psi_t)}{\dot{m}_f \psi_f}$

3 CASE STUDY

3.1 Vessel under study:

Our study will be based on the machinery arrangements of a typical container vessel; its main particulars are shown in Table 2.

Table 2: Ship understudy main particulars.

Length (m)	200	Propulsion engine	21,770 KW
Breadth (m)	29.8	Diesel generator set (KW)	3 sets: 1890 KW @ 720 rpm
Draught (m)	11.4	Boiler	2.5t/h @ 9 bars
Gross tonnage	26,190	Cargo capacity	2600 TEU

The ship machinery consists of the propulsion system and auxiliary machines.

- Propulsion engine: this ship is fitted with one main diesel engine (ME) for propulsion turning single propeller at a maximum speed of 25 Knots. The diesel engine is converting chemical energy to mechanical energy by burning fuel oil.
- Auxiliary diesel engine (AE): The ship is also fitted with three diesel generators of 1890 KW each producing electric power needed for auxiliary machinery operations and cargo reefer containers feed.
- Auxiliary machines necessary for the safe operations of propulsion system, navigation system and crew living condition on board: boilers, cooling pumps, lubricating pumps, fuel oil treatment, firefighting system, air compressors, air conditioning, deck machinery, etc.
- Auxiliary steam boiler (AB): One oil-fired boiler is installed onboard for heating systems of fuel oil, lubricating oil, cooling water, sanitary and air conditioning. During sea passage, the exhaust gases flow from the main engine is enough to produce steam without the need for firing the oil-fired boiler. In the port and during ship slow down, the main engine exhaust gas flow is missing or reduced therefore, the oil-fired boiler is used to supply the heating steam.

3.2 Measurements

The ship is trading between north Europe, Morocco and West Africa in the regular line. We collected the noon reports and ship's records for two round voyages lasting approximatively three months. Sorting of these data was necessary, as without this, our calculations could be altered because of the zero speed or consumption values corresponding to periods of stop or anchorage. We filter the data and ignore the non-consistent values recorded by mistake or inaccuracy by the crew.

In addition to the noon report, the engine and auxiliary engine performance records at different loads were provided by the crew in addition to equipment instruction books. Some parameter values were collected from engine logbook mainly engine parameters and electric power consumption.

Fig. 3 represents the ME load during the three-voyage period under study. Most of the time the engine is used below 60% of its nominal power for example ME was running during

3% of total running time at 5% of load only. In Table 3 we summarized the average percentage of load versus running hours. The fuel oil consumption at each % of load and the exhaust temperature are obtained from ship's records. The fuel oil chemical energy is obtained from eqn (6) while the exhaust gas flue is calculated based on the engine specifications and tests as stated in the engine approved technical files.

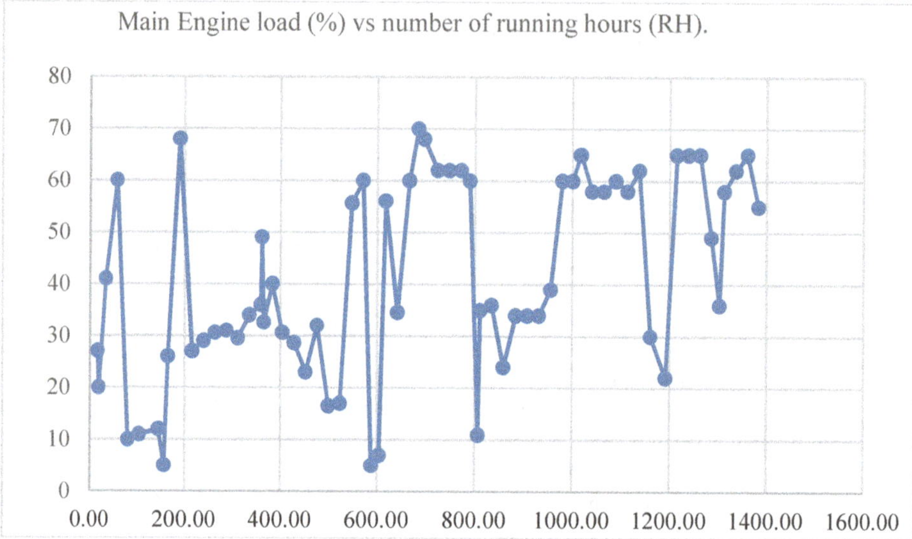

Figure 3: Engine running load over time.

Table 3: Main engine average load, power and fuel oil consumption vs % running hours.

Load %	5	15	25	35	55	65
% of total time	3	11	15	24	15	32
Power (KW)	1,090	3,100	5,100	7,150	11,700	14,050
FO cons. (Kg/h)	410	750	1,120	1,416	2,192	2,625
E f (MJ) (6)	4,795	8,771	13,109	16,560	25,634	30,698
Exhaust gas temp after T/C	300	295	290	290	280	260
Exhaust gas flow (t/h)	20	39	58	75	115	140

The same approach was adopted for the AE parameters as shown in Table 4. The average power is 1050KW, which is represents 55% of the engine nominal power. According to the performance record provided by the crew, at 55% load the diesel engine exhaust temperature after the turbine is 395°C (the energy and exergy values for AE shown in Table 6 are calculated based on the engine parameters at 55% of load). The boiler is used mainly when the main engine is stopped, to provide the steam needs in the port or at anchorage. The boiler is adjusted to 70% of maximum load. The parameters stated in Table 5 are taken from the ship's records and boiler specifications.

Table 4: Auxiliary engine load vs running hours.

AE (KW)	880	850	900	950	1,000	1,050	1,100
Period (h)	18	15	17.25	428	44.5	358	50.5
AE (KW)	1,150	1,200	1,250	1,300	1,350	1,400	1,500
Period (h)	72	24	772	45	45	12.5	40.5

Table 5: Auxiliary boiler parameters (values based on the steam diagram at 7 bars).

	Water	Steam	Fuel oil	Air	Combustion gas	Flue gas
Temperature (°C)	85	183.5	145	30	–	357
Flow rate (Kg/h)	1,050	1,050	71	1,189	1,260	1,260
Enthalpy (KJ/Kg)	285	2782	42,700	303.64	3,504	361
Entropy (KJ/Kg °K)	1.135	5.56	1.99	6.88	7.0	1.9

4 RESULTS AND DISCUSSION

4.1 Energy and exergy inputs

The fuel used onboard is the heavy fuel oil IFO 380 Density @ 15°C, kg/m^3. We take the net calorific values of fuel oil: NCV = 42.7 MJ/Kg and the exhaust gas specific heat Cp = 1.1KJ/Kg °K [17], where the values shown in Table 6 are calculated.

4.2 Efficiencies

By substitution of values shown in Table 6 in the eqns (4), (7)–(9) and (13)–(15), we calculate the different efficiencies shown in Table 7.

Table 6: Component calculated parameters.

	At sea			At port			Total		
	ME	AE	AB	ME	AE	AB	ME	AE	AB
FO cons. (ton)	2,313	380	–	–	105	29.28	2,313	485	29.28
Running hours	1,384	1,384	–	–	521	412	1,384	1,905	412
$\dot{E}_f$ (TJ)	98,765	16,226	–	–	4,483.5	1,272.5	98,765	20,709.5	1,272.5
$\dot{B}_f$ (TJ)	105,086	17,264	–	–	4,770.4	1,353.9	105,086	22,035	1,353.9
Cons. /hour	1.67	0.274	–	–	0.201	0.084	1.67	0.25	0.071
$\dot{E}_f$ (TJ/h)	71.36	11.72	–	–			71.36	10.87	3.09
$\dot{B}_f$ (TJ/h)	75.93	12.47	–	–			75.93	11.57	3.29
$\dot{E}_f$(%)	85.9	14.1	–	–	70.49	29.51	81.8	17.15	1.05
$\dot{B}_f$ (%)	85.9	14.1	–	–	70.49	29.51	81.8	17.15	1.05

Table 7: Diesel engine efficiencies.

	Av. (KW)	$\dot{Q}_{ex}$(KW) (7)	$\dot{Q}_d$ (KW) (4)	$\dot{B}_{ex}$ (KW) (13)	$\dot{B}_Q$(KW) (14)	γ (8) (%)	γ (15) (%)	Sfoc (9) g/KwH
ME	7,530	5,750	6,542	2,079	233	37	35.7	221
% of $\dot{E}_f$	38	29	33	-				
% of $\dot{B}_f$	35.7	-	-	11.5	1.1			
AE	1,050	960	1,009	243		34.5	32.7	238
% of $\dot{E}_f$	34.5	32	33.5	-	36			
% of $\dot{B}_f$	32.7	30	-	10	1.12			

For the total voyage, the propulsion (main engine) is responsible for 82% of the ship's total energy (fuel oil) consumption. Electric power production accounts for 17 to 18% of fuel energy demand, while the heating by auxiliary boiler is approximatively at 1%. This is because the heating is mainly maintained by ME exhaust gas boiler where exhaust gas wasted heat is recovered.

At sea, 82% of energy consumption is due to propulsion while in the port 70.5% of the energy consumed is for producing electric power needs.

In the main engine, 37% of the input energy flow (fuel oil) is converted to work (propulsion), 29% of energy is transferred to exhaust gas, part of this energy is recovered by exhaust gas economizer producing steam for heating needs. The remaining 34% is wasted to the environment through the cooling system, shaft frictions, and radiations. Part of the cooling heat is recovered by domestic heating system.

The specific fuel oil consumption measured is 221 g/KWh, much higher than the maker's value which is 180. This is due to low load, the ship operating at a low load of less than 55%, while the engine is designed for running at 85% of nominal power.

Less than 35% of fuel oil exergy is transformed to break power; 12% is evacuated by exhaust gas; the rest is lost by exergy destruction.

The auxiliary engine is a four-stroke type engine, 34% of the fuel consumed is converted to electric power, 32% is wasted to engine exhaust gas and 34% expelled to the cooling system, radiations, mechanical frictions.

The exergy balance is similar to the main engine: 33% of fuel energy is transformed to work while 10% is evacuated by exhaust gas and the remaining is destructed by heat transfer.

The specific fuel oil consumption measured is 238 g/KWh much higher than the maker's value which is 210. The reason is the same of the main engine, running at 55% load away from a design point. It was demonstrated by other literatures [18] that for such auxiliary engine size the global energy efficiency can be enhanced by installing economizer on the exhaust system similar to the one fitted on main engine. Hence the steam produced will partly compensate the lack of steam produced by main engine economizer in the port

This result can be extended to similar vessel's category however it should be noted that the container vessels are operated at low speed due to market shrinkage and bunker rate, in normal operations the share of propulsive power would be much higher.

The auxiliary engine is designed for normal safe operation of the ship and supplying the reefer containers. Due to low market the number of reefer containers loaded are less than the ship's capacity. Otherwise the auxiliary engine could be operated at higher loads and at efficient range (generally 75–85% of the maximum output).

By equations stated in Table 1 and parameters from Table 4, we calculate the boiler's efficiencies as shown in Table 8.

Table 8: Boiler efficiencies.

	Energy	Exergy
Combustion	$\eta_c = \frac{\dot{m}_g h_c}{\dot{m}_f h_f} = 1$	$\mu_c = \frac{\dot{m}_g.\psi_c}{\dot{m}_f.\psi_f} = 0.60$
Heat exchange	$\eta_h = \frac{\dot{m}_s(h_s - h_t)}{\dot{m}_f h_f} = 0.86$	$\mu_h = \frac{\dot{m}_s(\psi_s - \psi_t)}{\dot{m}_g(\psi_c - \psi_g)} = 0.65$
Boiler	$\eta = \frac{\dot{m}_s(h_c - h_t)}{\dot{m}_f h_f} = 0.86$	$\mu = \frac{\dot{m}_s(\psi_s - \psi_t)}{\dot{m}_f \psi_f} = 30$

The energy efficiency of the boiler is 86% which is within the range of marine boiler, the lost energy is partly expelled by exhaust gas and partly due to blowing down operation, surface heat transfer conditions. Most of the heat loss take place in the heat exchanger.

The overall exergy efficiency is approximately 30%, mostly due to irreversibility. The exhaust gas heat can be recovered from the flue gas through a heat exchanger which may be used for preheating boiler fuel oil or feed water.

5 CONCLUSION

In this study, we represented the energy and exergy balance of a typical container vessel. The study is based on actual measurements and records provided by the ship crew. It has been demonstrated that the propulsion plant is responsible for 82% of the total energy demand, electric power generation is accounted for 17% while the production of steam is responsible for 1% which is negligible compared to others. Steam needs are mainly supplied by an exhaust gas economizer fitted on the main engine.

The overall energy and exergy efficiencies of the ship are approximatively 34–36% and mainly influenced by diesel engine performances. The boiler efficiency is not affecting much the ship energy and exergy efficiencies.

The low efficiencies can be explained by low performance of diesel engine at low loads. The container vessels are designed for a cruise speed of 24 knots but due to the high cost of fuel oil, companies are adopting economic speed to reduce fuel consumption. The same has been observed for the ship understudy, in that the auxiliary engine is mostly operating at 60% of the nominal power, probably due to low fret. The generator is designed for the safe operation of the ship and reefer container energy supply.

There's a huge amount of energy lost by the cooling system and exhaust gas: up to 65% of total energy consumed.

There is considerable potential for waste heat recovery on container vessels. Various solutions are already existing and a plethora of technologies available for retrofitting on existing vessels however to improve the overall energy efficiency techno-economic study is to be done on waste heat recovery solutions for each category of the vessels and trading mode.

REFERENCES

[1] International Maritime Organization (IMO), Executive summary and final report: Third IMO GHG study 2014, 2015.

[2] IMO, Resolution MEPC.203(62): Amendments to the annex of the protocol of 1997 to amend the international convention for the prevention of pollution from ships, 2011. www.imo.org/en/KnowledgeCentre/IndexofIMOResolutions/Marine-Environment-Protection-Committee-(MEPC)/Documents.

[3] IMO, Resolution MEPC.278(70), Amendments to the annex of the protocol of 1997 to amend the international convention for the prevention of pollution from ships, 2017. www.imo.org/en/KnowledgeCentre/IndexofIMOResolutions/Marine-Environment-Protection-Committee-(MEPC)/Documents.

[4] Coraddu, A., Oneto, L., Baldi, F. & Anguita, D., Vessels fuel consumption: A data analytics perspective to sustainability. *Soft Computing for Sustainability Science, Studies in Fuzziness and Soft Computing Series*, ed. C. Cruz Corona, p. 358, 2016.

[5] Perera, L.P., Mo, B. & Kristjánsson, L.A., Identification of optimal trim configurations to improve energy efficiency in ships. *IFAC-PapersOnLine*, **48**(16), pp. 267–272, 2015.

[6] Lin, Y.H., Fang, M.C. & Yeung, R.W., The optimization of ship weather routing algorithm based on the composite influence of multi-dynamic elements. *Applied Ocean Research*, **43**, pp. 184–194, 2013.

[7] Cocak, G. & Durmusoglu Y., Energy efficiency analysis of a ship's central cooling system using variable speed pump. *Journal of Marine Engineering and Technology*, **17**(1), pp. 43–51, 2017.

[8] Aijjou, A., Bahatti, L. & Raihani, A., Enhanced ship energy efficiency by using marine box coolers. *Advances in Science, Technology and Engineering Systems Journal*, **3**(6), pp. 83–88, 2018.

[9] Zaili, Z. & Zhaofeng, Y., Exergy analysis of ship main propulsion plant integrated energy system. *Proceedings of the International Conference on Technology Management and Innovation*, 2010.

[10] Shi, W., Stapersma, D. & Grimmelius, H.T., Analysis of energy conversion in ship propulsion system in off-design operation conditions. *WIT Transactions on Ecology and Environment*, vol. 121, WIT Press: Southampton and Boston, 2009.

[11] DNVGL. World Fleet MACC 2030 ver 24. www.dnvgl.com/publications.

[12] Kotas, T.J., *The Exergy Method of Thermal Plant Analysis*, Elsevier: Amsterdam, 2013.

[13] Wilbur, C.T. & Wight, D.A., *Pounder's Marine Diesel Engines*, 6th ed., 1984.

[14] Abedin, M.J., Masjuki, H.H., Kalam, M.A., Sanjid, A., Rahman, S.M.A. & Masum, B.M., Energy balance of internal combustion engines using alternative fuels. *Renewable and Sustainable Energy Reviews*, **26**, pp. 20–33, 2013.

[15] Martyr, A.J. & Plint, M.A., The energy balance of an IC engine. *Engine Testing*, 4th ed., Elsevier, 2012.

[16] Kaushik, S.C. & Kumar Singh, O., Estimation of chemical exergy of solid, liquid and gaseous fuels used in thermal power plants. *Journal of Thermal Analysis Calorimetry*, **115**(1), pp. 903–908, 2014.

[17] Coskun, C., Oktay, Z. & Ilten, N., A new approach for simplifying the calculation of flue gas specific heat and specific exergy value depending on fuel composition. *Energy*, **34**(11), 2009.

[18] Henriksson, D. & Nyman Vierto, R., *A Case Study of Potential Savings that Can Be Made by Installing an Exhaust Gas Economizer on an Auxiliary Diesel Engine*, Chalmers University of Technology of Gothenburg: Gothenburg, 2016.

COMPARATIVE STUDY OF TWO TYPES OF MEDIUM TEMPERATURE PHASE CHANGE MATERIALS

SOL-CAROLINA COSTA[1], MOHAMMAD ISMAIL[1], MURAT KENISARIN[1], KHAMID MAHKAMOV[1], DAVID MULLEN[2], ELVEDIN HALIMIC[2] & KEVIN LYNN[2]
[1]Northumbria University, UK
[2]Aavid Thermacore, UK

ABSTRACT

This work presents a numerical comparative study of the melting and solidification processes of two medium temperature phase change materials (PCMs) for latent heat thermal energy storage (LHTES) applications. The PCMs compared are an inorganic solar salt and a Sn-based metal alloy, with the phase change temperatures around 220°C. The numerical comparative study was conducted in a 2D rectangular unit model with constant heat flux, the simulations were conducted deploying the enthalpy method in Ansys FLUENT software. The numerical results were validated using experimental results for one of the phase change materials studied. The charging and discharging times of the designed LHTES were studied for both phase change materials, finding that the metallic PCM LHTES showing superior performance. The charging and discharging time obtained for the solar salt LHTES, duplicate the time required using metallic PCM LHTES. Moreover, the final average temperature for the solar salt LHTES, shows significantly compare with the phase change temperature.

Keywords: PCM, LHTES, melting, solidification, metallic PCM, thermal properties.

1 INTRODUCTION

For several decades, latent thermal energy storage systems (LHTES) have been successfully used in numerous engineering applications such as solar plants, thermal management solutions, etc. In general, the phase change materials (PCMs) have a high energy storage density and small temperature variation during LHTES charging and discharging periods. The main disadvantage of currently used PCMs (organic materials and hydrated salts) is the low thermal conductivity which limits their application for processes that need high heat transfer rates. Therefore, it is necessary to deploy methods, which would enhance the heat transfer in the LHTES. There are several techniques used to enhance the heat transfer in the LHTES and these include adding high thermal conductive particles or placing high thermal conductivity insertions, such as metallic fins or porous matrices [1]. The performances of finned heat pipes coupled to different PCMs for cooling of electronic devices were experimentally and numerically studied in [2]. Metallic PCMs (MPCMs) were first considered for thermal storage more than four decades ago [3], however, due to mainly excessive weight and cost, compared to organics and salt hydrate PCMs, these were not used in LHTES applications and there is a very limited number of studies in this area. Disadvantages, related to the excessive weight, are partially compensated by the high thermal conductivity of appropriate metallic materials and the high thermal capacity of LHTES per unit of volume. The feasibility of application of MPCMs improves with the cost reduction of metallic materials using low purity alloys.

The aim of this work is to compare numerically the melting and solidification processes in two small physical models of medium temperature LHTES with solar salt and MPCM. The melting processes are studied in a rectangular shaped LHTES cells with the application of constant heat flux, provided by cylindrical heating cartridges, placed in the cells.

WIT Transactions on Ecology and the Environment, Vol 237, © 2019 WIT Press
www.witpress.com, ISSN 1743-3541 (on-line)
doi:10.2495/ESUS190041

2 MEDIUM TEMPERATURE PHASE CHANGE MATERIALS

The LHTES cells for experimental tests were developed with a thermal storage capacity of about 1 kWh to model medium temperature solar thermal storage applications at temperature levels of about 220°C. The thermal energy accumulated in a larger scale LHTES can be used, for example, to extend the operational period of an organic Rankine cycle (ORC) turbine [4] or other converters in solar thermal electricity plants [5].

Fig. 1 presents results from literature review of information on PCMs to be used in the temperature range from 300 to 1,200 K with indication of their costs per kWh [5]. In the medium temperature range (about 220°C) mainly salts and metals make up a significant fraction of available PCMs. However, only for a few of them the complete set of thermal properties is available in the literature [6]. Among materials, solar salt (60 wt.% $NaNO_3$+ 40 wt.% KNO_3) and the equimolar eutectic blend (46.7 wt.% $NaNO_3$+40 wt.% KNO_3) are those which were extensively investigated [7]. Also, tin and its alloys can be used as a PCM in special cases, in which the weight and cost are not critical factors for an application.

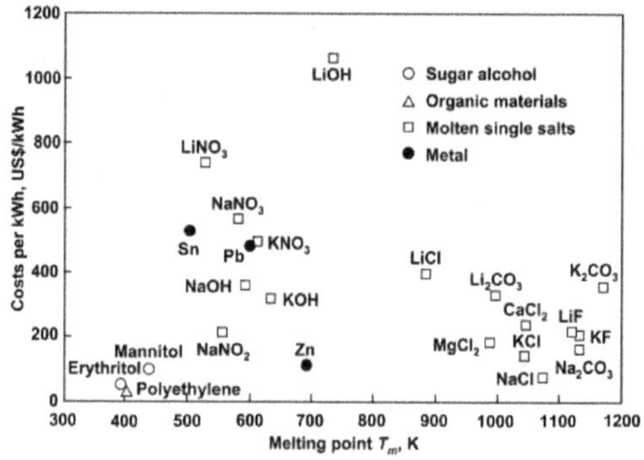

Figure 1: Media costs of PCMs [5].

In the temperature range studied, the preferable candidates are solar salt and tin based alloys (see Table 1). In this work, efforts were concentrated on obtaining the thermo-physical properties of three metallic PCMs, including their phase transition temperature, the heat of fusion and the specific heat.

Fig. 1 shows the costs per kWh for different PCMs and around the 500 K temperature level application Sn, pure nitrate salts KNO_3 and $NaNO_3$ have similar values. In 2005 the price of pure tin was lower than 10.000 US$/tonne. The price for pure tin has increased in the last years; however, the price depends on its purity grade and in certain cases the price difference between high and low purity materials is significant. When evaluating the cost of PCM/additive composition it is vital to include the cost of additives, used to enhance the thermal conductivity of composite, since quite often the cost of additives per unit of mass are significantly higher than that for pure PCMs.

2.1 Thermo-physical properties of the selected PCMs

The melting temperature, the heat of fusion and the specific heat capacity were determined for solar salt, Sn, 96.5Sn-3.5Ag (wt.%) and 99.3Sn-0.7Cu (wt.%) using Differential Scanning Calorimeter (DSC) Setaram EVO-131. The results of the DSC carried out in the present work are shown in Fig. 2.

Table 1: Physical properties of PCM candidates.

Wt.%	**60 wt.% $NaNO_3$ +40 wt.% KNO_3**	**Tin (Sn)**	**96.5Sn-3.5Ag**	**99.3Sn-0.7Cu**
H_m (kJ/kg)	91.5 [8]	64.0 (PW)	62.3 (PW)	64.5 (PW)
$T_{solidus,\ m}$ (°C)	218.9 [8]	232.2 (PW)	221.4 (PW)	226.8 (PW)
$T_{liquidus,\ m}$ (°C)	**225.0 [8]**	**237.2 (PW)**	**227.7 (PW)**	**232.9 (PW)**
H_s (kJ/kg)	87.7 [8]	60.7 (PW)	62.3 (PW)	64.6 (PW)
$T_{solidus,\ s}$ (°C)	212.4 [8]	170.6 (PW)	214.6 (PW)	201.2 (PW)
$T_{liquidus,\ s}$ (°C)	218.5 [8]	173.0 (PW)	212.4 (PW)	–
ρ_l (kg/m^3)	1,958 [9]	6974 [10]	6,998 [11]	7,067 [12]
β (1/K)	3.47E-04 [9]	1.14E-04 [10]	9.55E-05 [11]	1.13E-04 [12]
k (W/m K)	0.380 [9]	73(s) [6]	84 (s) [13]	58.9 (s) 44.8 (l) [14]
Cp (J/kg.K)	1620 [9]	222 [6]	260 [15]	220 [6]

PW = present work.

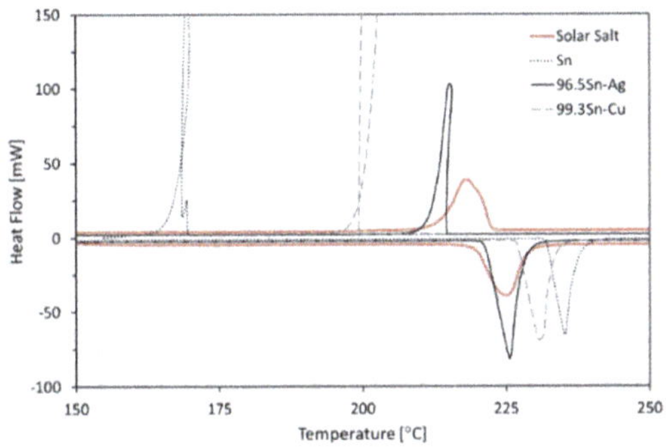

Figure 2: The typical DSC analysis for the PCMs.

It can be observed that the melting temperatures are between 225 and 237°C for these PCMs. However, during the solidification process of pure tin, a high degree of supercooling (approximated 50°C) is observed during the transformation from the liquid to solid state. The supercooling is clearly reduced for the Sn-based alloys, however, the alloy which has the phase transition temperatures closer to that of solar salt is 96.5Sn-3.5Ag (wt.%). In general, the thermal hysteresis can be observed during phase change in PCMs.

Table 1 summarizes the results obtained for metallic PCMs and solar salt. Each value is obtained as the average value of three measurements. This data obtained from DSC tests is used in the process of designing and numerical simulations of operation of the LHTES cells. For simulation purposes, data on the density, thermal conductivity and viscosity were obtained from other references.

When comparing the thermophysical properties, the main advantage of solar salt can be seen, which is the high heat of fusion. Its main drawback is the low thermal conductivity. The metallic PCMs have much higher values of the thermal conductivity but the heat of fusion per unit of mass is relatively small, resulting in the excessive mass of the LHTES.

2.2 The LHTES cell design

The LHTES cell is designed in order to provide thermal energy at the rate of 0.25 kW for the duration of 4-hour period [4]. In order to store this amount of the thermal energy as a latent heat, about 39 kg of solar salt or 58 kg of the metal alloy are necessary. The charging of the cell is carried out by applying a constant heat flux produced by cylindrical cartridges placed in the cell, this mean that the heat transfer rate to the LHTES is fixed. These cartridges simulate the charging process by heat pipes [4]. The number of heat pipes installed in the cell is two with the maximum heating/cooling rate of 125 W each with a length of 500 mm.

Table 2 shows the LHTES cells characteristics in terms of weight and volume. The mass of the PCM is increased by a factor of 1.5 using metallic PCMs and the volume is decreased by 60%. The volume of the solar salt LHTES cell is 0.020 m^3.

Table 2: LHTES cell characteristics.

PCM	Storage capacity energy (kWh)	Mass (kg)	Volume (m^3)
60 wt.% $NaNO_3$ +40 wt.% KNO_3	1	39	0.020
96.5Sn-3.5Ag (wt.%)	1	58	0.008
96.5Sn-3.5Ag (wt.%)	2.5	142	0.020

3 NUMERICAL MODELING OF LHTES

The aim of the numerical modelling is to compare the performance of these two different PCMs for application in medium temperature solar LHTES. The numerical modelling uses a 2D simplified representation of a cell considering the heat pipe as a constant heat flux device [16]. The physical geometry of the cell unit with heat pipes embedded is shown in Fig. 3. The width and height of the selected frontal area are 142 and 283 mm, respectively, for the 0.020 m^3 LHTES and 58 and 283 mm, respectively, for the 0.008 m^3 LHTES.

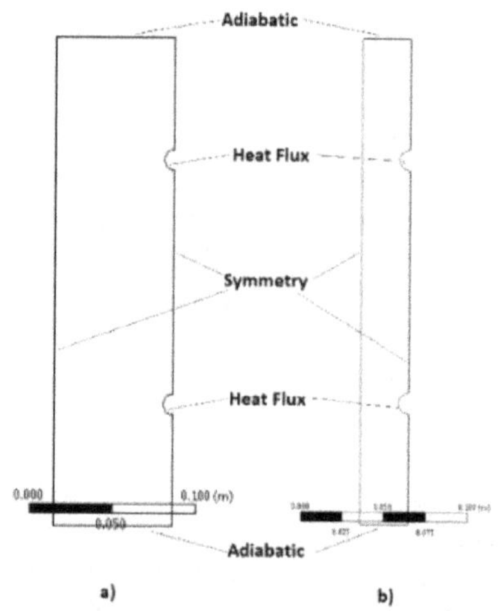

Figure 3: The 2D geometry of the cell. (a) Solar salt (0.020 m^3); and (b) Metallic (0.08 m^3).

The commercial software ANSYS Fluent 19.1 was used for simulations of the melting and solidification processes, see Fig. 3. The enthalpy-porosity technique [17]–[19] is deployed to model the PCM melting and solidification processes. In this technique, the mushy zones of PCM are considered as porous media and the porosity of each element in the mushy zone is assumed to be equal to the liquid fraction of that element.

The power law differentiating scheme and SIMPLE method for pressure–velocity coupling are used to solve the momentum and energy conservation equations. The PRESTO scheme is adopted for the pressure correction equation. In the present study, the time step was set to 0.2 s and 0.15 s for the 125 and 250 W of heat transfer rate, respectively. The mesh sensibility analysis is evaluated at 125 W by comparing the variations in liquid fraction and average temperature over three different grid sizes (78,600, 52,400 and 36,800 elements) as is shown in Fig. 4. The results demonstrate that there are not significant differences between the two first grids studied, therefore the computational grid size selected is Mesh 2 (Fig. 4) which has 52,400 elements.

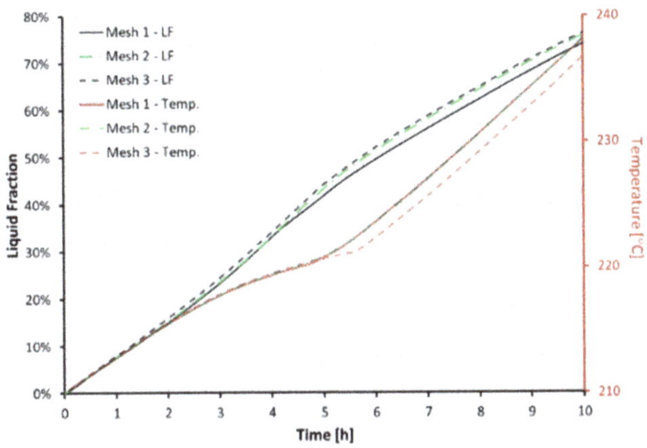

Figure 4: Mesh size sensibility analysis.

The assumptions made in the modelling process are as follows:

- Flow is 2D, laminar and incompressible (no temperature gradient along the LHTES);
- The properties of the PCM are considered constant in both the solid and the liquid phases. The used thermo-physical properties of PCM are listed in Table 1;
- The natural convection effects are modelled by the Boussinesq approximation;
- Thermal expansion is disregarded;
- Constant and uniform heat flux on the surface of heat pipes calculated, considering that the heat transfer rate capacity for both heat pipe is 125 and 250 W;
- Thermal contact resistances are ignored at the interfaces;
- In the cell unit, all vertical walls are symmetry planes and the upper and bottom boundaries are adiabatic walls (Fig. 3).

The numerical model presented in this work for the solar salt was previously experimentally validated using two identical (in the design) prototypes (Fig. 5). The melting and solidification experiments were carried out to validate the numerical melting/solidification models and to study the heat transfer mechanisms in the LHTES [8].

Figure 5: Top view of the container with the PCM [8].

Table 3 shows the experimental and numerical results on determination of the melting and solidification times in both prototypes studied. The liquid fraction, formed in simulations, is also listed for corresponding elapsed times. It can be observed that for the melting process numerical and experimental results agree well with the accuracy within 7%. However, the numerical and experimental results for solidification processes deviate by 18%. The main reason for such the difference is that the solidification process is a free cooling process taking place overnight (governed by the heat losses) and the rate of heat extraction is not controlled during the experiments.

Table 3: Experimental and numerical results [8].

	Melting TC-4 = 225°C	Solidification TC-4 = 212°C
Steel experimental	4.7 h	4.2 h
Steel numerical	4.4 h (LF = 82%)	4.5 h (LF = 11%)
Al. exp.	3.7 h	4.0 h
Al. num.	3.8 h (LF = 97%)	4.8 h (LF = 0%)

3.1 Numerical results for charging process

The initial temperature for modelling the charging process is set to 210°C for both PCMs. At this temperature PCMs are in the solid phase at the start of simulations. Fig. 6 shows the evolution of the liquid fraction during the melting process for solar salt and the metallic PCM at heat transfer rate of 125 W. It can be observed that the melting processes of these PCMs are totally different. In the case of solar salt, the buoyancy effect is observed from the start of the melting process. For the metallic PCM, the thermal conduction dominates during all the melting process, which starts much later compared to the case with solar salt. Such behaviour is expected because the Prandtl number of the metallic alloys is close to 0 ($Pr \ll 1$) and therefore the thermal diffusivity dominates, in other words, the heat transfer due to conduction is much greater compared to the convection mechanism. In the case of solar salt, the Prandtl number is approximated 24 ($Pr \gg 1$) therefore the momentum diffusivity dominates and the convection mechanism plays much more intensive role in transferring energy in comparison to pure conduction [20]. In LHTES, where the melting and the solidification of the PCM alternate, the melting process is limited by thermal conductivity and natural convection, and the speed of the solidification process is mainly defined by the thermal conductivity [5].

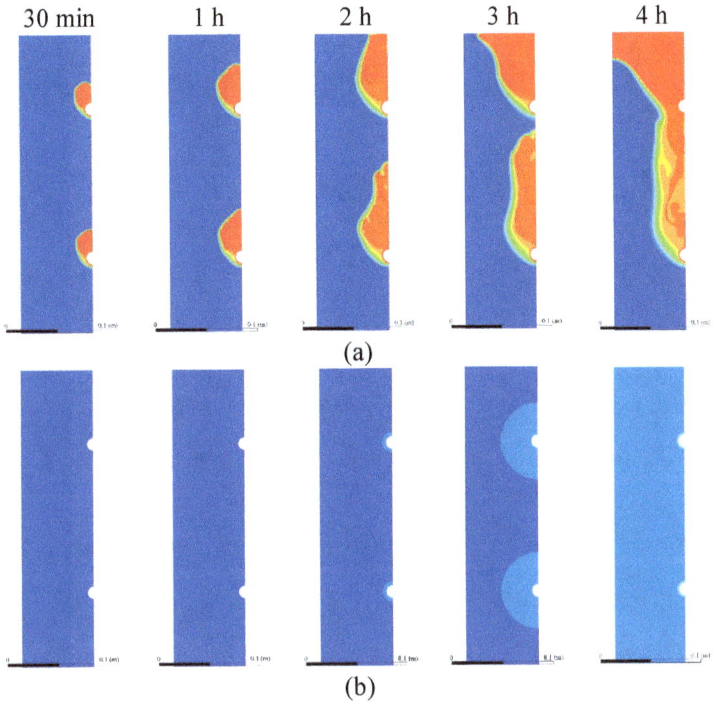

Figure 6: Liquid fraction contours for melting process (125 W) at different elapsed times. (a) Solar salt; and (b) Metallic PCM.

Fig. 7 shows the liquid fraction versus the time during the melting process. In the case of the metallic PCM the melting process is dominated by the thermal diffusivity, therefore the final melting time can be easily calculated (the charging time at the heat transfer rate imposed). Fig. 8 shows the average temperature versus time during the melting process. The phase change during the melting process in the metallic PCM can be divided into three stages: sensible heating of the solid phase, latent heating and sensible heating of the liquid phase. For solar salt, the melting process is dominated by the natural convection at the initial stage and the melting rate is close to that of the metallic PCM for the same storage capacity. After all the material above the heat pipes turn into liquid phase, the melting process rate is drastically reduced, and the thermal diffusivity dominates the melting process. Another important difference is that the temperature in the LHTES with solar salt increases significantly before the melting process is completed (see Fig. 8), therefore the heat thermal storage occurs simultaneously in the form of sensible and latent heat. During the melting process of solar salt, at the initial stage the melting rate is dominated by natural convection therefore, the impact of the heat pipes arrangement is more important than in the case of metallic PCMs. Consequently, the study of the more efficient geometrical shape and heat pipes arrangement is of paramount importance in order to maximize the effect of natural convection.

Table 4 summarises the numerical results on determination of the melting times and average temperature for both types of PCMs studied. The times are the elapsed time from the beginning of the simulation until the liquid fraction makes 90% of the volume. It is observed

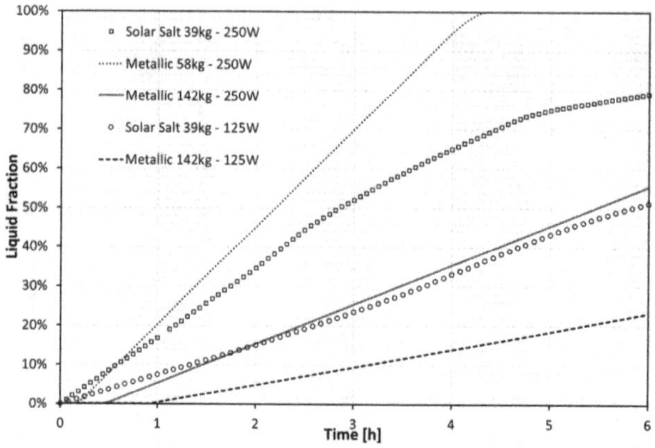

Figure 7: Liquid fraction versus time during melting process.

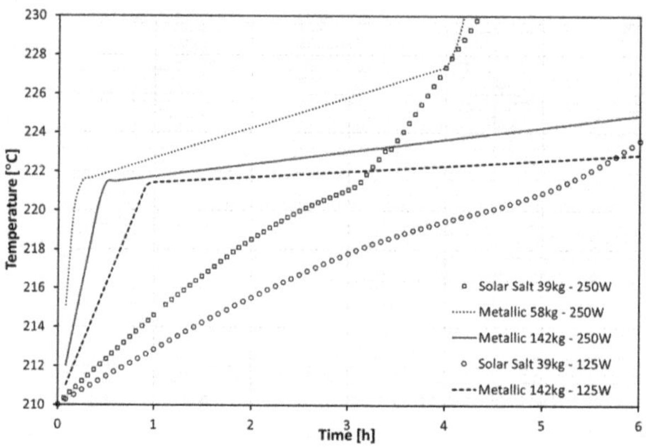

Figure 8: Average temperature versus time during melting process.

that for the metallic PCMs the average temperature in the LHTES is practically the "liquidus" temperature in the different geometries and power supply values studied. However, for solar salt, the temperatures display a significant deviation from the "liquidus" temperature. In Table 4 it can be observed that the metallic LHTES, designed for 1 kWh heat capacity, can be charged in approximated 4 hours (heat transfer rate) and the LHTES with solar salt with the same heat storage capacity require approximately 8 hours. In all the cases the heat transfer rate is controlled but depending on the nature of the heat thermal storage process the melting rate is totally different because the thermal storage process is not purely based on latent heat.

3.2 Numerical results for discharging process

The initial temperature for modelling the solidification process is set to the "liquidus" temperature during the melting for solar salt and metallic PCM, which is 225 and 227.7°C, respectively. In the LHTES the initial discharging temperature is determined by a number of

Table 4: Numerical results for melting process (90% liquid fraction).

PCM/LHTES capacity	Total HP power (W)	Melting time (h)	Average temperature (°C)
Solar salt/1kWh	250	~8.0	~270
Metallic/1 kWh	250	3.8	227
Metallic/2.5 kWh	250	9.5	227
Solar salt/1kWh	125	15.7	270
Metallic/2.5 kWh	125	19.5	227

factors: the final charging level, the thermal losses, the characteristic of the PCMs, etc. Fig. 9 shows evolutions of the liquid fraction during the solidification process for solar salt and metallic PCM at heat transfer rate of 125 W. It can be observed that the solidification process for the metallic PCM is practically homogeneous in the whole volume of the LHTES. For solar salt there is a clear stratification process at the beginning and then the solidified material gradually grows circularly around the heat transfer devices.

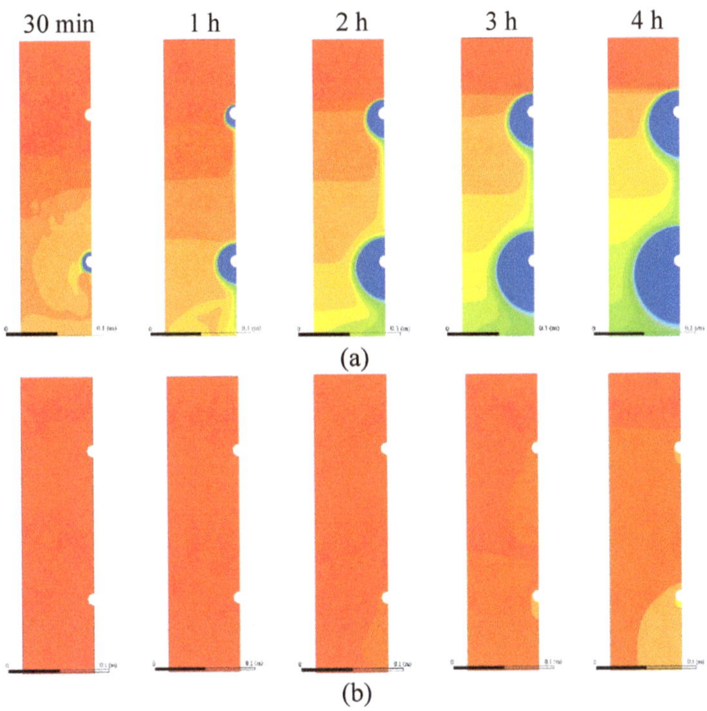

Figure 9: Liquid fraction contours for solidification process (heating rate of 125 W) at different elapsed times. (a) Solar salt; and (b) Metallic PCM.

Fig. 10 shows the liquid fraction versus the time during the solidification process for the considered PCMs at two different heating rates. The metallic PCM solidification process is almost linear as in the melting process. Fig. 11 shows the average temperature versus the time during the solidification process and the phase change process for the metallic PCM can be clearly identified as it was during the melting process. For solar salt the solidification process has predominantly a linear tendency, however the solidification rate changes at the instance when only 30% of volume if filled by the liquid fraction. The average temperature in the LHTES with solar salt decreases continuously along of the solidification process.

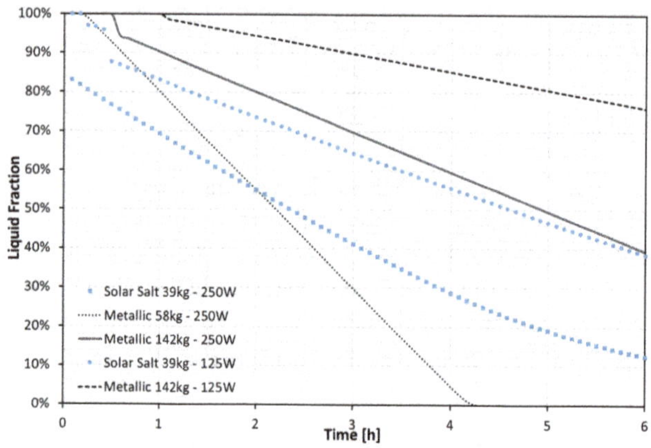

Figure 10: Liquid fraction versus time during solidification process.

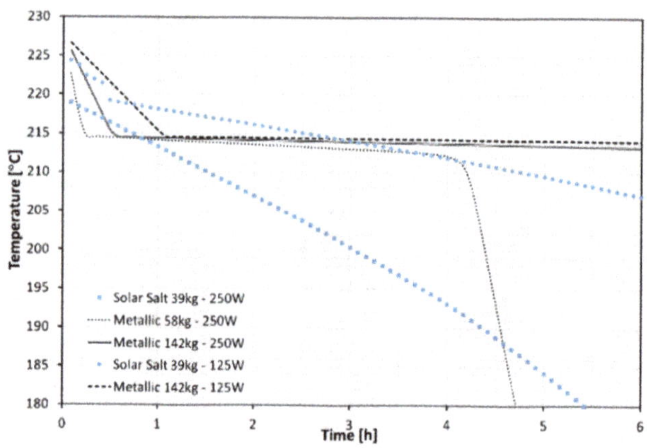

Figure 11: Average temperature versus time during solidification process.

Table 5 shows the numerical results for the solidification times and average temperature for both PCMs studied. The times are the elapsed time from the beginning of the simulation until the volume occupied by the liquid fraction is 10% of the total volume. It can be observed that for the metallic PCMs the average temperature in the LHTES is practically the "solidus" temperature in all cases. However, for solar salt, the temperatures present a significant deviation from the "solidus" temperature. It can be also seen that for the metallic PCM the final solidification time is practically the same as the melting time. Results show that the metallic LHTES, designed for 1 kWh capacity can be discharged in approximately 4 hour time providing heat at the rate of 0.25 kW. However, the solar salt LHTES designed for 1 kWh capacity requires approximated 8 hour time to be discharged for the same conditions, and this represents a significant deviation from the initial requirement. Considering the final average temperature of solar salt after discharging, the LHTES will require additional energy and time to be recharged.

Table 5: Numerical results for solidification process (10% liquid fraction).

PCM/LHTES capacity	Total HP power (W)	Solidification time (h)	Average temperature (°C)
Solar salt/1kWh	250	6.7	166
Metallic/1 kWh	250	3.8	213
Metallic/2.5 kWh	250	8.9	213
Solar salt/1kWh	125	11.1	187
Metallic/2.5 kWh	125	20.3	213

The numerical results indicate that there are significant differences in the performance of the LHTES using solar salt or metallic PCMs. The differences are in terms of discharging and charging times, average temperatures, melting and solidification processes. Moreover, the design of the LHTES also presents significant differences in terms of dimensions, weight, shape, simplicity and cost. The design of the solar salt LHTES is more complex because the numerical analysis is required in order to evaluate the best arrangements and configuration of heat transfer devices inside the unit. Furthermore, the use of thermal conductivity enhancement methods is necessary in order to fulfil charging and discharging time requirements, affecting the final weight and cost of the LHTES.

4 CONCLUSIONS

In the present work, the melting and solidification processes of two medium temperature PCMs, namely solar salt and metal alloys, were studied in a rectangular LHTES cell. The melting process in the case of the metallic PCM is dominated by the thermal conduction, and in the case of solar salt at the beginning of the process by the natural convection and then, for the thermal conductivity. Therefore, the design of the metallic LHTES with a constant heat flux type heater/sink can be carried out with relatively simple analytical calculations, which also provide good estimation of the charging and discharging times. The design of the solar salt LHTES is more demanding because the multidimensional numerical analysis might be required in order to determine the best arrangement and configuration of heat transfer devices. The results shown superior performance of the metallic PCMs in terms of controlling the charging and discharging rates. However, the applications of this kind of PCMs are limited by the increment in weight and cost.

ACKNOWLEDGEMENTS

This project has received funding from the European Union's Horizon 2020 Research and Innovation Programme under Marie Sklodowska-Curie grant agreement No. 705944 (THERMOSTALL). This study is also a part of the "Innova MicroSolar" Project, funded by the European Union's Horizon 2020 Research and Innovation Programme under the grant agreement No 723596.

REFERENCES

[1] Lin, Y., Jia, Y., Alva, G. & Fang, G., Review on thermal conductivity enhancement, thermal properties and applications of phase change materials in thermal energy storage. *Renewable and Sustainable Energy Reviews,* **82**, pp. 2730–2742, 2018.

[2] Yang, X.-H., Tan, S.-C., He, Z.-Z. & Liu, J., Finned heat pipe assisted low melting point metal PCM heat sink against extremely high power thermal shock. *Energy Conversion and Management,* **160**, pp. 467–476, 2018.

[3] Birchenall, C.E.T., Thermal storage in metals. *Sharing the Sun: Solar Technology in the Seventies, Proceedings of the Joint Conference*, Winnipeg, Canada and Cape Canaveral, Florida, International Solar Energy Society, **8**(A77-48910 23-44), pp. 138–154, 1976.

[4] Mahkamov, K. et al., Development of a small solar thermal power plant for heat and power supply to domestic and small business building. *ASME Power & Energy*, Lake Buena Vista, Florida, USA, p. 10, 2018.

[5] Hoshi, A., Mills, D.R., Bittar, A. & Saitoh, T.S., Screening of high melting point phase change materials (PCM) in solar thermal concentrating technology based on CLFR. *Solar Energy,* **79**(3), pp. 332–339, 2005.

[6] Jankowski, N.R. & McCluskey, F.P., A review of phase change materials for vehicle component thermal buffering. *Applied Energy,* **113**, pp. 1525–1561, 2014.

[7] Kenisarin, M. & Mahkamov, K., Salt hydrates as latent heat storage materials: Thermophysical properties and costs. *Solar Energy Materials and Solar Cells,* **145**(3), pp. 255–286, 2016.

[8] Costa, S.C. et al., Experimental and numerical study on melting of solar salt in a finned metallic container. Presented at *ASME 2018 International Mechanical Engineering Congress and Exposition*, American Society of Mechanical Engineers, 2018.

[9] Bauer, T., Pfleger, N., Breidenbach, N., Eck, M., Laing, D. & Kaesche, S., Material aspects of solar salt for sensible heat storage. *Applied Energy,* **111**, pp. 1114–1119, 2013.

[10] Mitra, S.S., Viscosity and density of liquid metals. *Physica,* **24**(1–5), pp. 155–156, 1958.

[11] Yagodin, D., Sidorov, V., Janickovic, D. & Svec, P., Density studies of liquid alloys SnAg and SnZn with near eutectic compositions. *Journal of Non-Crystalline Solids,* **358**(21), pp. 2935–2937, 2012.

[12] Dobosz, A. & Gancarz, T., Reference data for the density, viscosity, and surface tension of liquid Al–Zn, Ag–Sn, Bi–Sn, Cu–Sn, and Sn–Zn eutectic alloys. *Journal of Physical and Chemical Reference Data,* **47**(1), 2018.

[13] Meydaneri, F., Saat, B. & Ozdemir, M., Thermal, electrical, microstructure and microhardness properties of the eutectic Sn-3.5wt.%Ag alloy, 2013.

[14] Ocak, Y., Aksöz, S., Maraşlı, N. & Keşlioğlu, K., Thermal conductivity and interfacial energies of solid Sn solution in the Sn–Ag–In ternary alloy. *Chemical Physics Letters,* **496**(4–6), pp. 263–269, 2010.

[15] Çadırlı, E., Şahin, M., Kayalı, R., Arı, M. & Durmuş, S., Dependence of electrical and thermal conductivity on temperature in directionally solidified Sn–3.5 wt.% Ag eutectic alloy. *Journal of Materials Science: Materials in Electronics,* **22**(11), pp. 1709–1714, 2011.

[16] Tiari, S., Qiu, S. & Mahdavi, M., Discharging process of a finned heat pipe-assisted thermal energy storage system with high temperature phase change material. *Energy Conversion and Management,* **118**, pp. 426–437, 2016.

[17] Tiari, S., Qiu, S. & Mahdavi, M., Numerical study of finned heat pipe-assisted thermal energy storage system with high temperature phase change material. *Energy Conversion and Management,* **89**, pp. 833–842, 2015.
[18] Lohrasbi, S., Miry, S.Z., Gorji-Bandpy, M. & Ganji, D.D., Performance enhancement of finned heat pipe assisted latent heat thermal energy storage system in the presence of nano-enhanced H_2O as phase change material. *International Journal of Hydrogen Energy,* **42**(10), pp. 6526–6546, 2017.
[19] Guo, C. & Zhang, W., Numerical simulation and parametric study on new type of high temperature latent heat thermal energy storage system. *Energy Conversion and Management,* **49**(5), pp. 919–927, 2008.
[20] Cengel, Y.A., *Heat Transfer*, 2nd ed., McGraw-Hill: London, 2007.

ENERGY AND SUSTAINABLE PERFORMANCE OF A MULTIFUNCTIONAL FAÇADE

INÊS SIMÕES[1], NUNO SIMÕES[1,2], INÊS SANTOS[1], MICHAEL BRETT[1], SÉRGIO TADEU[1] & HÉLDER SILVA[3]
[1]ITeCons – Instituto de Investigação e Desenvolvimento Tecnológico para a Construção, Energia, Ambiente e Sustentabilidade, Portugal
[2]Department of Civil Engineering, Faculty of Sciences and Technology of the University of Coimbra, Portugal
[3]Silva & Ventura, Lda., Portugal

ABSTRACT

Concerns about increasing comfort levels, reducing energy consumption, reducing waste and using natural, recyclable and reusable products is forcing the construction industry to invest in research and the development of innovative solutions that address these goals. There is still a great market demand for having glazing architectural solutions without neglecting all the other functional characteristics. In this context, a sustainable kit solution for façades was developed, consisting of a modular system that allows the combination of several functionalities and configurations. The system integrates the use of expanded cork (Insulation Cork Boards – ICB) and photovoltaic (PV) glass fastened to the support structure (wall) using an innovative fixation system which allows the disuse of frame profiles, resulting in façades with an elegant appearance. This system feature has both architectural and technical integration advantages. The first part of the paper presents the system's physical behaviour when subjected to real climate conditions. The study was performed on a prototype installed and monitored in the central region of Portugal. In addition, and to better quantify the benefits of having a solution that combines a low thermal resistance envelope using natural thermal insulation and PV energy production, a life cycle of the full solution is performed.

Keywords: photovoltaic façade, PV's energy production, life cycle.

1 INTRODUCTION

The buildings sector, in the European Union, currently accounts for approximately 40% of final energy consumption and therefore a significant proportion of total CO_2 emissions. This percentage has been increasing, surpassing industry and transport [1]. Following the commitments made under the Kyoto Protocol, a set of policies and measures have been established by the EU, including improving the energy performance of buildings (Directive 2002/91/EC [2] and Directive 2010/31/EU [3]).

To reach this goal, it is necessary to improve the thermal behaviour of a buildings envelope, setting appropriate energy performance requirements for technical building systems, for future buildings and in the requalification of existing buildings [4]. In addition, the concept of Nearly Zero Energy Buildings (NZEB), which is now seen as a realistic solution for reducing energy consumption and CO2 emissions, has to be considered [5]. Therefore, it is imperative to promote energy efficiency using new constructive technologies, which contribute to the limitation of energy consumption and which are additionally environmentally sustainable due to their material and production processes. This new reality requires the introduction of a "holistic design" that combines different components and technologies in a building through an integrated approach, to the detriment of a traditional dispersed approach of specific or individual elements [6] and that takes into account the whole life cycle. The life-cycle analysis (LCA) is a fundamental tool in the decision-making process for reducing environmental impact, since it focuses on assessing the environmental effect of the processes or products and can be applied successfully at different scales [7].

www.witpress.com, ISSN 1743-3541 (on-line)
doi:10.2495/ESUS190051

The concept of "Almost Zero Energy" in buildings makes evident the increasing importance of the surrounding energy performance. To the classical principles of the "skin" conception of the buildings of Utilitas (function), Firmitas (solidity, construction) and Venustas (beauty, form) [8], adds the principle of Energy. In fact, the "skin" or envelope of the building has an important function in the energetic balance of the building, because this element is a boundary where the external energy interacts with the inner environment [9]. As such, the promotion of the energetic efficiency of the building's envelope involves the use of "passive" constructive technologies (for instance, externally insulated application solutions), and "active" solutions (for example, energy from renewable energies) [6]. One of the latest lines of envelope development is the integration of multiple façade functions. The trend will be to integrate not only windows, solar protection and other passive devices, but also active systems such as cabling infrastructure, lighting, decentralized air conditioning systems, automatic solar protection, photovoltaic cells, and others. The integration of active and passive systems into multifunctional façades opens a new chapter of potential capital gains that guides the energy efficiency of buildings to a new era [10]. Looking at the different categories of multifunctional façades, it is necessary to emphasize the energy collector façade, which can aggregate, for example, the functions of climate protection, passive solar capture, natural light capture, reception and pre-air conditioning of new air, with an additional function of active energy production, namely using solar thermal collectors or photovoltaic cells.

It is in this context that this article presents a façade solution with thermal and energetic characteristics, optimised according to the climatic requirements of different locations in the European territory, incorporating a cork insulation layer (ICB) and a hidden interface of photovoltaic glass fixation to the wall.

The system development included numerical simulations that were carried out to assess the performance of the solution in different locations. This development took into account the life cycle analysis in a comparative way (with competitive reference solutions) of the system. In order to demonstrate the energy efficiency of the solution, real-time monitoring of a large-scale prototype was carried out. These results are presented below.

2 SYSTEM DESCRIPTION

The developed solution is a sustainable kit solution for façades which consists in a modular system that is able to perform multiple functions and present multiple configurations. The system integrates insulation Cork Boards – ICB and photovoltaic (PV) glass fastened to the support structure (wall) using an innovative fixing system which will allow the disuse of frame profiles, resulting in façades with an elegant appearance (slimframe) (Fig. 1). The system uses wire ropes for a fast and easy application.

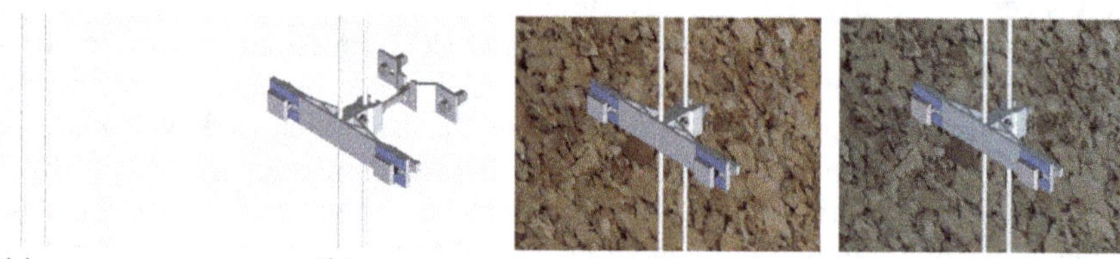

(a) (b) (c) (d)

Figure 1: 3D drawing with the assembly sequence of the system. (a) Cable (wire rope) alignment; (b) Installation of the fastening system; (c) Application of insulation cork boards; and (d) Installation of the glasses.

This system was designed to be applied on buildings up to 30 m high and is suitable for new and rehabilitation projects. The fastening system enables a quick assembly system and allows the easy replacement of glasses.

Some laboratory tests were carried out in order to assess the mechanical performance of the system, such as resistance to wind load test, air permeability test and water-tightness test. After that, the system's energetic behaviour when subjected to real climate conditions was evaluated using a prototype installed and monitored in the central region of Portugal. In addition, and in order to quantify the other benefits of the solution, a life cycle and economic assessment of the full solution was also performed. These results are presented below.

3 REAL ENERGY PRODUTION OF THE SYSTEAM

After the development of the system, a large-scale test specimen was constructed, using the following elements:

- 15 semitransparency photovoltaic panels (10% of transparency), with amorphous silicon cells. The dimension of each panel was 1245x635 mm2;
- air chamber of 30 mm;
- 40 mm thick insulation Cork Boards;
- air chamber of 100 mm;
- 18 mm thick wooden board.

The test specimen was subjected to real climatic conditions (Fig. 2). The system was placed in order for the photovoltaic panels to have south orientation. A ventilation system was placed at the top end of the rear side of the wall. This system is activated when the temperature between the system air chamber and the ambient temperature is equal to or greater than 15°C.

The system is being monitored so as to register the temperatures (at several locations), the solar energy production and the period in which the ventilation is activated. Fig. 3 shows the location of the thermocouples and ventilation system. The system has been monitored since 3rd March.

Figure 2: Test specimen.

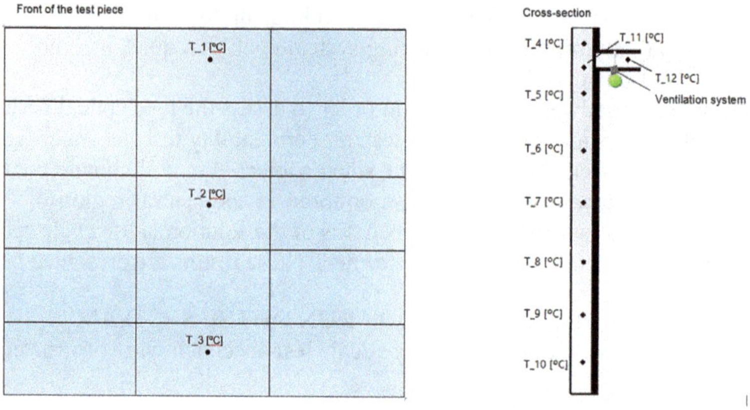

Figure 3: Drawing of the test specimen with the location of the thermocouples and ventilation system.

Fig. 4 shows the results obtained during a period of two weeks of measurement: one week unfavourable to photovoltaic systems and another week of clear days. It is observed that the ambient temperature was similar to the temperature at the base of the air box of the system. This figure shows that the ambient temperature was close to the temperature at the base of the air chamber (thermocouple T_10). The temperature difference between the base and top of the air chamber was, sometimes, greater than 20°C.

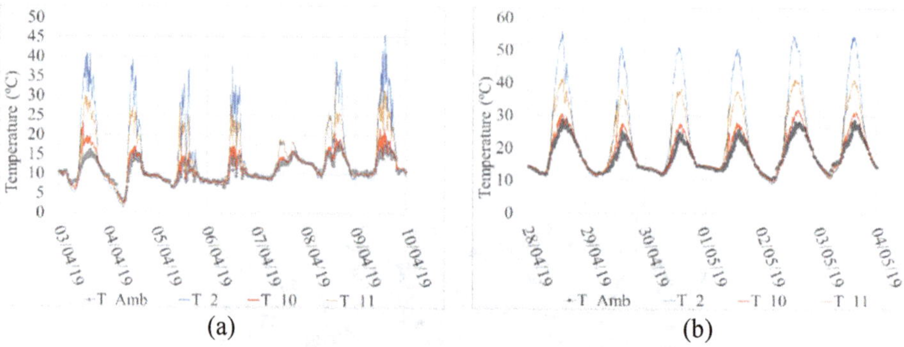

Figure 4: Experimental results. (a) Unfavourable week to photovoltaic systems: 3rd–10th April; and (b) Favourable week to photovoltaic systems: 28th April–4th May.

Fig. 5 presents the temperature at the top of air chamber, the ambient temperature and power generation on 7th and 28th April, while Fig. 6 presents the power generation with solar radiation, on these days. The 7th April was a cold and cloudy day. On that day, the ventilation system did not switch on during the whole day. The power generation was approximately 3.64 Wh/panel (4.53 Wh/m2). The 28th May was a clear sky day (Fig. 5). The ventilation system was on for more than 12 h. The system produced about 163 Wh/panel (206.9 Wh/m2). If we assume that all PV panels produce the same and also that every day of a month has the same energy production, in one month the system generated about 76 kWh. This value is higher than the simulated using numerical models.

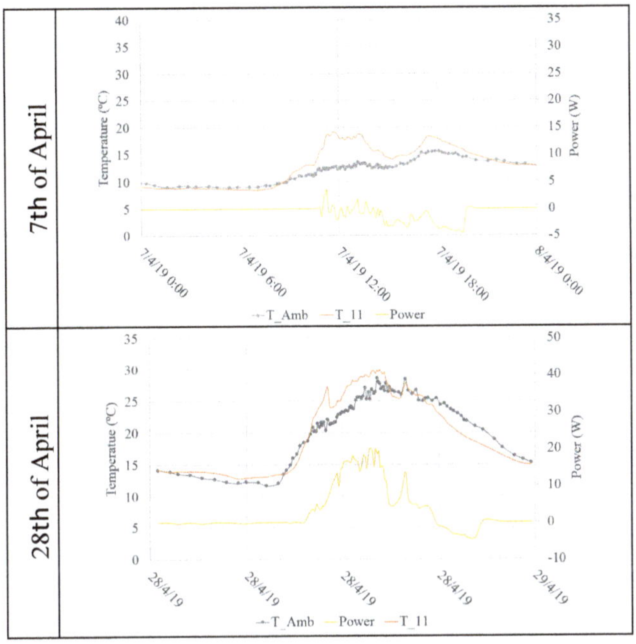

Figure 5: Experimental results on 7th and 28th of April: temperature and power generation.

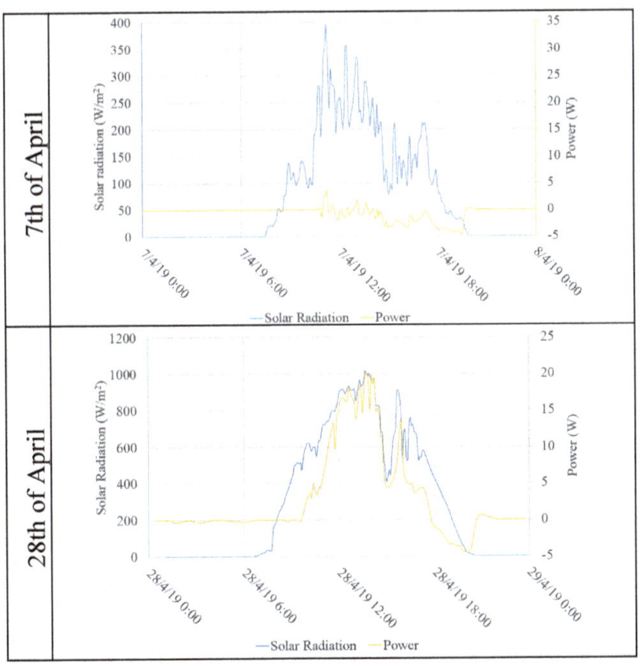

Figure 6: Experimental results on 7th and 28th of April: power generation and solar radiation.

4 LIFE CYCLE ANALYSIS

The innovation of this construction system is the fastening system. However, it is in our interest to characterise the performance of the system as a whole. For this reason, at a first stage, only the life cycle of the fixation system was analysed, and at a second stage, it was extended to the system as a whole.

4.1 Life cycle analysis

The life cycle analysis (LCA) had the objective of quantifying the environmental performance of the multifunctional façade system integrating cork and photovoltaic system with occult fixation developed and described above, as well as to identify the elements of inventory with greater contribution to the different categories of impact. The study was divided into two parts:

1. The product stage of the innovative fixation system (#PFSlim) was analysed, and its performance was compared with that of a conventional spider fixation (#PFConv) – Fig. 7;
2. The façade system, including the innovative fixation, was analysed, taking into account three variations in the glass unit: (2.1) façade system without photovoltaic (#SFSPV); (2.2) façade system with photovoltaic 4% transparency (#SFPV4%); (2.3) façade system with photovoltaic 10% transparency (#SFPV10%). Additionally, three building locations were considered (case 1: Munich, case 2 Oslo, and case 3: Lisbon).

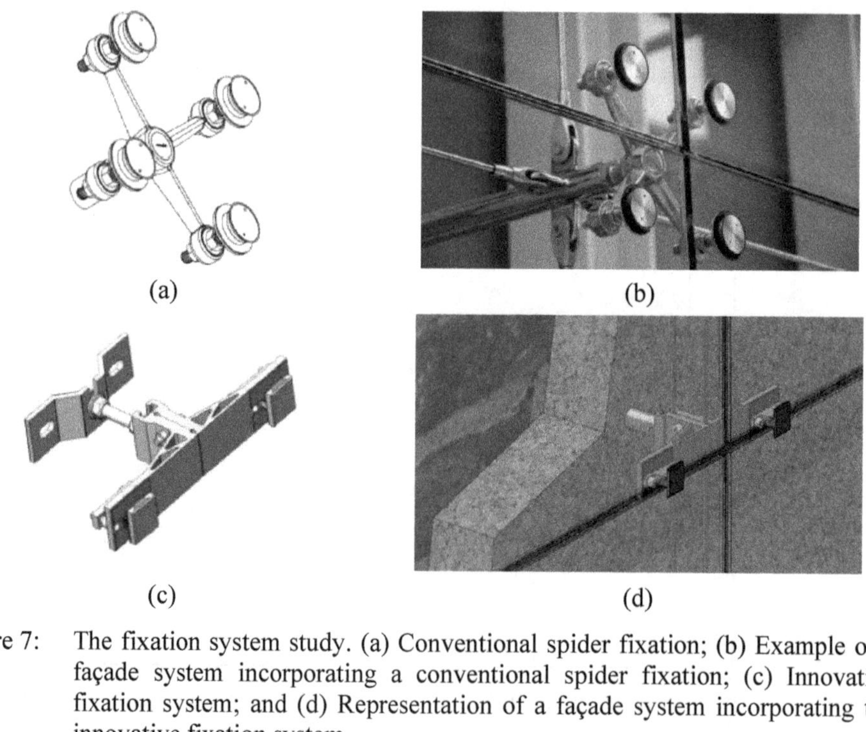

Figure 7: The fixation system study. (a) Conventional spider fixation; (b) Example of a façade system incorporating a conventional spider fixation; (c) Innovative fixation system; and (d) Representation of a façade system incorporating the innovative fixation system.

The development of the lifecycle model of the innovative fixation system took into account the stages of extraction and processing of raw materials, transport and production of the components. The LCA study of the façade system took into account the stages of extraction and processing of raw materials, transportation, component production, construction of façade systems and a scenario of use, namely electric energy production and heating/cooling needs. A façade system with capacity to produce electricity through photovoltaic panels is a multifunctional system. According to standard ISO 14044:2006/Amd1:2017, for multifunctional systems, the allocation of environmental loads should be avoided by dividing the system into two or more processes or by expanding the boundaries of the system. In this study, to evaluate the benefits of producing and using solar energy it was necessary to expand the system boundaries, through the approach of avoided burdens. In this regard, it was considered that the production of solar energy replaces the electricity production of the grid, being equivalent. In this way, the production of solar energy is assumed as credit in the impacts of the life cycle analysis of the façade systems. The functional unit is 1 m^2 of multifunctional façade, including cork, glass and the fixation system, with a reference life of 30 years.

4.2 Analysis methodology

The study presented here uses the methodology of the life cycle analysis based on international standards ISO 14040:2006 and ISO 14044:2006/Amd1:2017. This methodology allows the quantification and evaluation of potential environmental impacts throughout the life cycle of a product, from the extraction of the raw materials to the end of its life, through the compilation of inflows and outflows (mass and energy flows). The CML-IA version 4.2 (Leiden University, April 2013) with a midpoint approach, was used for the evaluation of life-cycle impacts (LCIA). The following impact categories were considered: Abiotic Depletion – elements (ADP – elements), Abiotic Depletion – Fossil Resources (ADP – fossil resources), Global Warming (GWP), Ozone Layer Depletion (ODP), Photochemical Oxidation (POCP), Acidification of Water Resources and Soil (AP), Eutrophication (EP).

In this study, all data concerning the real production of fastening parts and façade systems were considered. The grid electricity production, considered in the usage scenario, was determined based on the Portuguese, Germany and Norway's production mix in 2014 (Electricity low voltage), available in the *Ecoinvent v3.4* database. The inventory data are in accordance with the quantities used in the production of fasteners, in the production of photovoltaic panels, in the construction of the prototype and in the calculation of a use scenario. To complement missing data or to compare and validate existing data, we also used bibliographic sources and the *Ecoinvent v 3.4* database.

4.3 Studied scenarios

The heating/cooling needs and the production of solar energy used in this study were published by Melo [11], obtained for a service building with an interior area of 20 m^2, which integrates a façade of 12 m^2. 80% of this façade is glazing (PAE = 80%). In this stage, the energy consumption was considered to meet the heating and cooling needs of the building for 30 years, as well as the production of solar energy by the photovoltaic system (when applicable). Table 1 resume the annual energy flows for the façade systems and localities under analysis.

Table 1: Annual energy flows for the façade systems and localities.

	#SFSPV	#SFPV4%	#SFPV10%
Case 1: Munich			
Cooling energy requirements (kWh/year)	210	0	60
Heating energy needs (kWh/year)	815	1050	915
Production of solar energy (kWh/year)	0	80	55
Case 2: Oslo			
Cooling energy requirements (kWh/year)	125	0	25
Heating energy needs (kWh/year)	1100	1290	1190
Production of solar energy (kWh/year)	0	60	45
Case 3: Lisbon			
Cooling energy requirements (kWh/year)	1010	220	500
Heating energy needs (kWh/year)	80	95	100
Production of solar energy (kWh/year)	0	150	110

4.4 Results

Fig. 8 shows the comparative environmental assessment of the product step of the fixation system. The results show that the #PFConv fastening have the worst performance for all impact categories, which is due to the use of stainless steel in most of the components and because of the use of material in greater quantity.

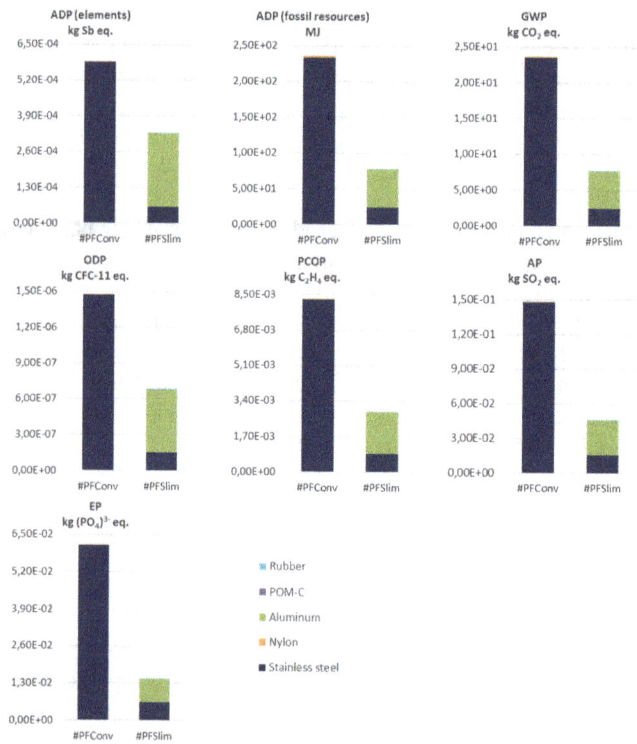

Figure 8: Environmental assessment of the product step of the fixation system.

Fig. 9 presents a comparative environmental assessment of the product stage of façade systems #SFSPV, # SFPV4% and # SFPV10%, for the environmental impact categories under analysis. It was observed that for the façade system #SFPV4% and #SFPV10%, the photovoltaic panels presented greater contribution to all the impact categories. This is because the production process of photovoltaic panels uses raw materials with a significant environment impact (mainly due to the lass, photovoltaic cells and encapsulation, which representing between 41.9 and 76.6% of the impacts). For this reason, the comparative evaluation between the façade systems shows that the #SFPV4% system performs worse in all impact categories, due to the productive process of the photovoltaic panels.

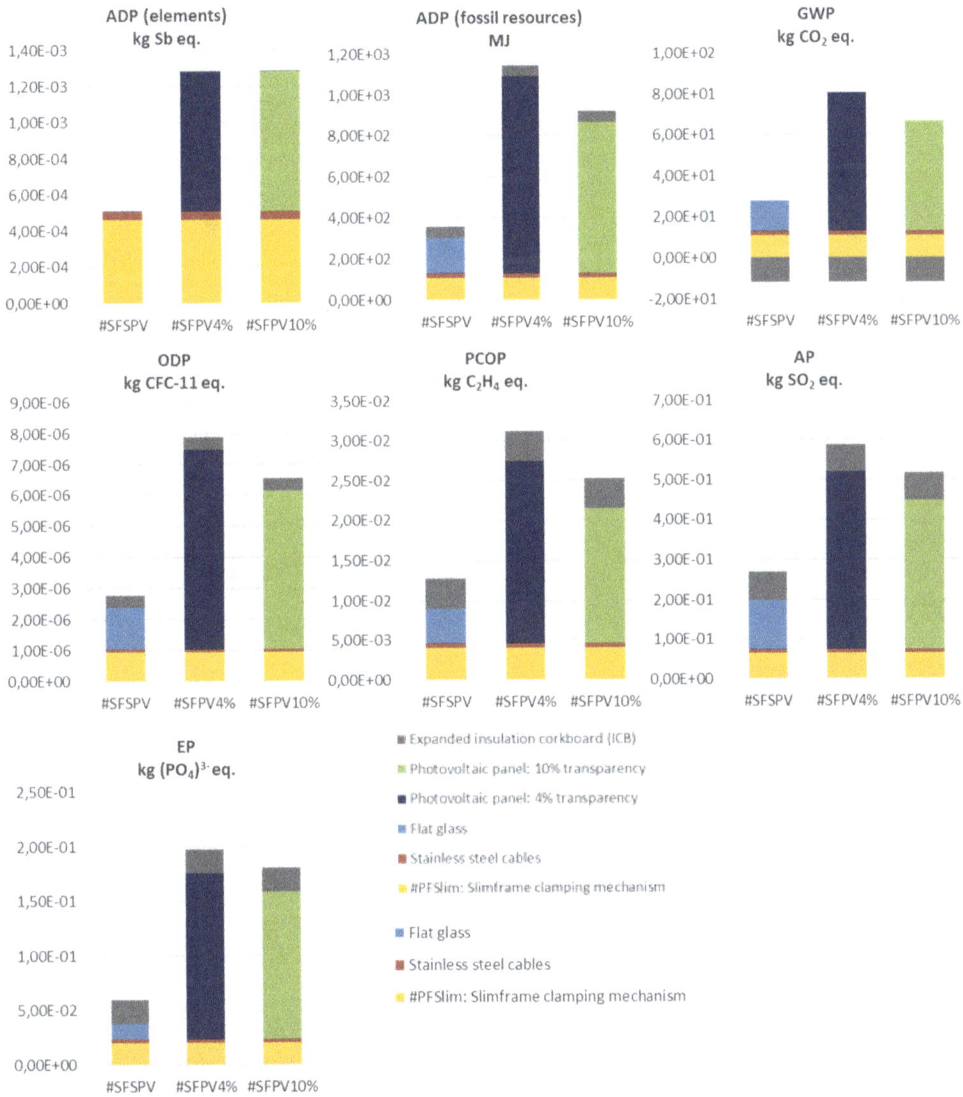

Figure 9: Comparative environmental assessment of the product stage of façade systems for the environmental impact categories under analysis.

Fig. 10 shows a comparative environmental assessment of each stage of life cycle and use scenario of façade systems under analysis and avoided impacts. The comparative evaluation shows that the location in Oslo presents better performance for all categories of impact, except for the OPD category, which is due to the energy mix of this country being composed by a large percentage of renewable energy. On the other hand, Munich's location exhibits the worst performance for all impact categories, except for the POCP category, which is due to the energy mix of this country being composed of a large percentage of non-renewable energy (e.g. coal), thus increasing impacts on the use stage. For Lisbon's location, the façade system #SFPV4% shows the best performance for all categories, except for the category ADP (elements), because this city is located in the Mediterranean, with mild winters, summers warmer and more sun exposed.

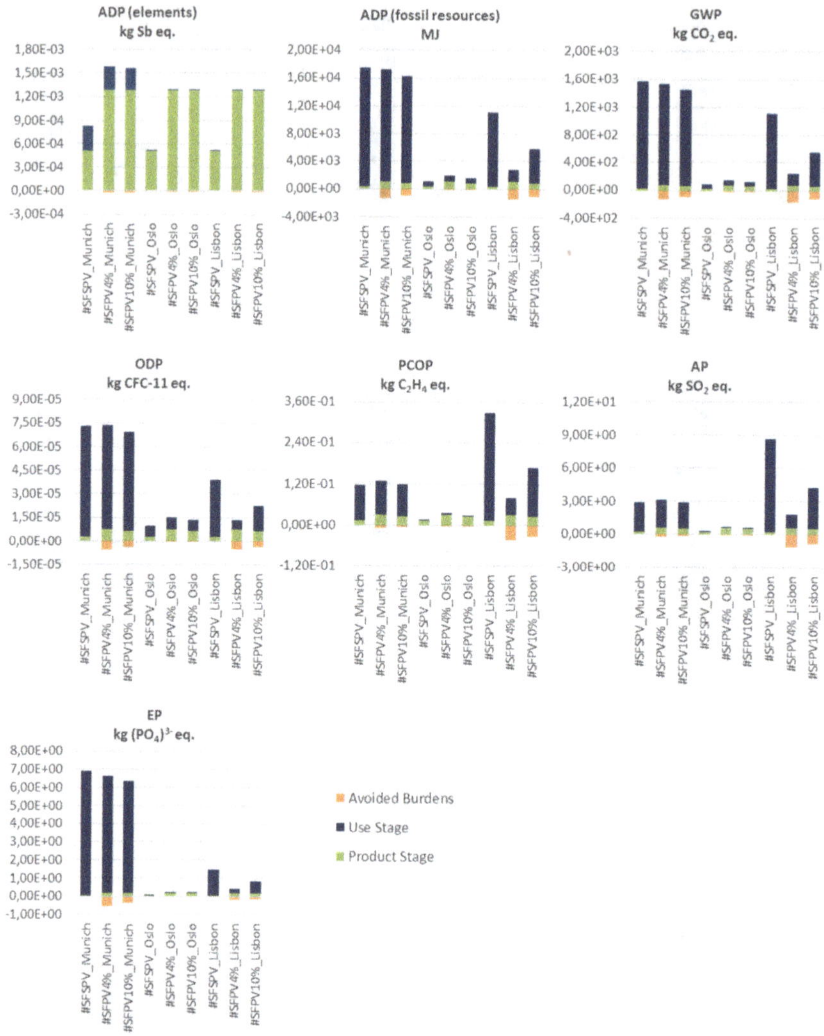

Figure 10: Comparative environmental assessment of each stage of life cycle and use cases of façade systems under analysis and avoided impacts.

It should be noted that in the use stage of the #SFPV10% and # SFPV4% façades, impact was avoided due to the replacement of the electric power of the grid by solar electric energy in the heating / cooling of the building.

The highest percentage of impact avoided was in Lisbon, because it was the city which had the highest production of solar energy.

5 CONCLUSION

This article presents a façade solution with thermal and energetic characteristics, optimised according to the climatic requirements of different locations in the European space, incorporating a cork insulation layer (ICB) and a hidden interface of photovoltaic glass fixation to the wall.

After system development, a real test specimen was placed outside a building and subjected to real climatic conditions. This system was monitored to monitor the temperatures and the energy production. The results show that the energy production was similar or even better than those obtained in the numerical simulation.

The life cycle analysis of the system was presented in a comparative way, and the results have confirmed environmental benefits.

ACKNOWLEDGEMENT

This work was framed within the CENTRO-01-0247-FEDER-033896 (S&V SunRoof) Project funded by Portugal 2020 through the COMPETE 2020.

REFERENCES

[1] Pérez-Lombard, L., Ortiz, J. & Pout, C., A review on buildings energy consumption information. *Energy and Buildings*, **40**(3), pp. 394–398, 2008.

[2] E. Commission e C. Document, Directive 2002/91/EC of the European Parliament and of the Council of 16 December 2002 on the energy performance of buildings, Off. J. Eur. Union, pp. 65–71, 2002.

[3] E. Commission e C. Document, Directive 2010/31/EU of the European Parliament and of the Council of 19 May 2010 on the energy performance of buildings, Off. J. Eur. Union, pp. 13–35, 2010.

[4] Ascione, F., De Rossi, F. & Vanoli, G.P., Energy retrofit of historical buildings: Theoretical and experimental investigations for the modelling of reliable performance scenarios. *Energy and Buildings*, **43**(8), pp. 1925–1936, 2011.

[5] Marszal, A.J. et al., Zero Energy Building – A review of definitions and calculation methodologies. *Energy and Buildings*, **43**(4), pp. 971–979, 2011.

[6] World Business Council for Sustainable Development (WBCSD), Energy Efficiency in Buildings Facts & Trends – Full report, 2008.

[7] Ortiz, O., Castells, F. & Sonnemann, G., Sustainability in the construction industry: A review of recent developments based on LCA. *Construction and Building Materials*, **23**(1), pp. 28–39, 2009.

[8] Vitruvio Pollio, M., *Vitrúvio – Tratado de Arquitectura*, Lisboa: IST Press, 2006.

[9] Bradshaw, V., *Building control systems*. New York: John Wiley and Sons, Inc., 1993.

[10] Castrillón, R.D.A., Integration of Active and Passive Systems in Glass Façades. Presented at *8th International Conference on Sustainable Energy Technologies,* Aachen, Germany, pp. 1–8, 2009.

[11] Melo, L., *Avaliação de desempenho de um sistema passivo de aquecimento integrando vidro fotovoltaico*, 2016.

ENERGY-RELATED HOUSING RENOVATIONS FROM AN EVERYDAY LIFE PERSPECTIVE: LEARNING FROM PORTUGUESE HOMEOWNERS

MARIA ISABEL ABREU, RUI A. F. DE OLIVEIRA & JORGE LOPES
School of Technology and Management, Polytechnic Institute of Bragança, Portugal

ABSTRACT

Changing the attitudes of existing building users towards sustainability is a paramount strategy towards a low-carbon society. One of the challenges lies in encouraging owners to renovate their houses. Recent studies have pointed out the relevance of the homeowner's social context which, more often than not, outstrips public awareness campaigns and even techno-economic advice. The decision to renovate seems to emerge from deep social issues rather than exclusively from the simple analysis of investment return or environmental awareness. The present study aims to bring discussion to these social subjects for single-family buildings in Portugal and identify key factors which may influence decision making. Firstly, a set of in-depth interviews were conducted with a group of experts. Subsequently, and based on the interviews, a group of Portuguese single-family homeowners, who undertook at least one energy-related piece of work, were surveyed. Together with the survey a structured observation method was put into practice. The results revealed that homeowner's stage of life, personal/family events, everyday practices, lifestyles and social network framework have a major influence on their prime intention to renovate their homes. The desire to maintain or change the routines and the way family uses the indoor spaces are common drivers, however, motivations linked to aesthetics, trends and social status were revealed to be extremely influencing. The decision process is negotiated inside the family involving not only desired domestic life conditions but also skills, understandings and social values shared by the dwellers. The energy-related renovations take place, in many cases, as a consequence of a previous decision to renovate, taken due to other reasons. It does not consist of one single event but is rather part of a practice of a "little-by-little renovation". Motivational policies should be drawn considering the energy-related renovations as any other type of renovation and not separately from the others.

Keywords: energy-related renovation, energy policy, houseowners, energy efficiency, decision-making, everyday life, social practice, drivers, residential building stock, Portugal.

1 INTRODUCTION

The housing stock represents 74% of the European building stock and is mostly inhabited by the owners: 70% in the European Union; 67% in the United Kingdom; 89% in Denmark and 74.7% in Portugal [1]. There is no doubt that reduction of CO_2 emissions from owner-occupied buildings is of major importance. In Portugal, the greatest potential for savings is in residential and services. This sector alone represented 42% of the contribution to primary energy saving goals of the total predicted for 2016 [2]. The implemented policies have failed to result in an execution rate of improvement procedures in the existing residential buildings that enables meeting the required targets [3], [4]. First of all, we must highlight that the common premise commonly used to support the arguments motivating homeowners has been that they wish to renovate in order to save energy and money, but are prevented from doing so mostly by financial constraints and by doubts regarding energy saving, financial return and the reliability of the professionals who will execute the works. These bases have been the support to behavioural sciences (and consequently to national policies) used to explain human behaviour regarding buildings energy efficiency. They have been mostly anchored in microeconomic issues, in the dynamics of owners' uptake of technology [5], on consumer behaviour and marketing [6]. Behavioural science has treated homeowners as individuals

www.witpress.com, ISSN 1743-3541 (on-line)
doi:10.2495/ESUS190061

who make rational choices in order to obtain certain results which are dissociated from home life [7]. This representation of homeowners' decisions to achieve goals and isolated from other family decisions is part of a speech centred on individual choice [8]. Therefore, recent approaches to the problem call for social science in order to conceptualise the study of decision-making regarding the renovation of the building stock as a process emerging from home life circumstances and subjected to several levels of influence [9]–[12]. This study aims to get an in-depth knowledge about energy-related renovation of single-family owner-occupied houses as part of the daily and the social sharing practices which constitute Portuguese family life.

2 CONTEXTUAL BACKGROUND ABOUT BEHAVIOURAL SCIENCE

Behavioural science has been based on the assumption that it is necessary to identify barriers and motivations and that these will account for homeowners' decision to renovate housing buildings with energy efficiency. The formal models used to understand this process have most often emphasised the advantages of energy efficiency and the financial issue. Personal and contextual influences have been used to explain the reasons behind homeowners' low uptake. This analysis targets decision-makers and not differentiated dwellers, energy efficiency and not indoor spaces improvements, the extraordinary and not everyday activity involving shared objects, skills and understandings, one-time renovation and not the "little-by-little renovation" and buildings and not homes with meaningful spaces [7]. Behavioural sciences are limited to the decisions rather than the process preceding them or the domestic context from which they emerge [7]. Nair et al. [13] suggested that there could be other factors that influence energy efficiency investment decisions such as lifestyles. Judson and Maller [10] also claim that policies only focusing on technical intervention issues have a limited scope. This kind of renovation cannot continue to be seen as a mere exercise of integrating technology into buildings. Stern [14] had previously identified the lack of interdisciplinarity in the studies about this subject and Lutzenhiser [11] criticises the restrictive model which has been used in policy-making regarding buildings' energy saving as it leaves out contributions from social science. Recently, Tjørring and Gausset [15] gave a reinforcement on this issue suggesting that social aspects often weighted more than economic aspects.

3 SOCIAL SCIENCE IN ENERGY RESEARCH: RECENT EVIDENCE

Energy use has a social nature [6]. Energy provides useful services allowing the functioning of normal social activities which are part of individuals' home life [16] and such practices are reproduced throughout time and space [17]. Comfort and convenience have become, according to current patterns, ordinary demands which impact on energy use [18].

The roles of social science in interdisciplinary energy research consists not only in dealing with social complexities encountered in pursuing aims driven mostly by engineering but also assisting in framing priorities, questions and options [19]. There are a representative number of studies that introduced social science to the investigation on housing renovation to reduce energy consumption considering homeowners' daily life [20]–[27]. The studies suggest that such practices surpass environmental concerns and the reduction of energy consumption. Renovating a building is an action which is negotiated within the family in response to the performance of family practices [10] where households' skills and home meanings have a specific role [17], [25]. Palm et al. [28] advise that energy consumption in households with more than one member is a collective rather than individualized process and call for in-depth analysis of household couplings. Wilson et al. [7] go further and claim that encompassing the decision-making process within home life conditions brings those deepest influences which are at the core of decision-making processes. They argue that renovating is oftentimes a way for dwellers to solve conflicts and tensions, achieve goals, take up opportunities and give wings to personal and family aspirations. The decisions to renovate are processes and such processes emerge from and take place within the daily home life circumstances (sometimes also triggered by extraordinary events). In many cases, the decision to adopt energy efficiency measures emerges later in the process to renovate others parts [29].

Homeowners' level of involvement in the renovation process (oftentimes doing some of the tasks themselves) also constitutes one of the reasons pointed out for starting the process. The renovation becomes part of their daily life, making them feel emotionally more attached to the house and often leading to touches of innovation [30].

In 2014, Gram-Hanssen [24] exposed other very important issue. Aesthetical and functional issues come first in the list of Danish homeowners. An example of this is the remodelling of kitchens and bathrooms, which continue to be the top preferences, while not leading to any investment return. These renovations are part of social norms regarding an image of prestige. Visible renovations are prioritized and frequently made before other non-visible measures although they know they are not economically reasonable [15]. Trends, status and lifestyles are factors this author also refers as boosters to renovation. Bravo et al. [31] demonstrated that renovation works are normally linked to aesthetic aspects of the house.

Wilson et al. [7] went further and identified levels of influence in the decision-making. The ultimate influences are at the core of the initial intention (e.g. why renovate?) and behavioural science is failing to work with them. Gram-Hanssen [24] mentioned the following example: if the family decide to change the windows or the roof, calculations on the energy saved as well as on the investment return may be decisive to choose the thickness of thermal insulation or the technical framing and glass solution. However, as far as the decision to renovate is concerned, investment return is rarely an influence.

Finally, Wilson et al. [7] consider that policy-makers should acknowledge that renovations in housing buildings are predominantly an improvement and adaptation of homes indoor spaces so as to better execute home life tasks and practices. They should incentivize and support the energy efficiency measures included in such renovations. The authors also recommend that researchers identify specific conditions of home life which are related to renovation activities. The homeowners may be encouraged to adopt energy efficiency measures in connection with these renovations even because the adoption of energy efficiency measures is often cost-effective during a general renovation [32].

4 MATERIALS AND METHODS

The analysis reported in this paper is based on a preliminary stage of a larger scale research project about the home-related practices and social perspective of energy-related renovation of private homes. On a preliminary phase and to cope with the aims of this study, a qualitative method was considered fundamental to in-depth knowledge, identifying social conditions and motivations and explaining them in more detail. A minor quantitative analysis complements the qualitative research results. A presential survey was made in 2016–2017 to a group of 57 Portuguese homeowners of single-family houses (included the homeowners' partners that were at home when the survey was made). Given the number of respondents, the results could be, eventually, not demonstrative of all cases and are limited to the universe of the circumstances found in the survey. Previously, three in-depth interviews to experts (informants) were also made to help to prepare the survey structure (the long-time experience was the main criteria). The survey was also tested in three homeowners to guarantee if all the important aspects were covered in order to answer the investigation questions and to see if the questions were clear. They were invited to put some comments about confusions, problems or recommendations. Based on this feedback, the final version of the survey was put into action at the home of the respondents to get a personal contact with them and their family. This option allowed face-to-face interaction, capture of both affective and cognitive aspects and gave the researcher the opportunity to clarify or explain the question for better responses. Therefore, it was implemented a qualitative research method involving structured behavioural observation – observational survey. The questions were both open-ended and closed-ended. The number and type of energy-related renovations made and the motivation factors where the central topics explored. The interviewer took notes about homeowners and their partners reactions to the questions using observation guides (parameters to be observed were clearly pre-defined) in order to gather data on individual behaviour or interactions between participants (facial expressions, oral comments and positive and negative stimuli).

The sample were selected, initially, consulting the municipality services to identify districts with single-family building typologies constructed between 1970 and 2000 (buildings more than 15 years old). Three urban municipalities in the North of Portugal (severe Winter Climatic Zone) were considered for a preliminary approach. The homeowners were first contacted personally at their home and then was scheduled a meeting. Knowing if they had already performed at least one type of energy-related work was the question made to the homeowners in the initial contact. Only the ones in this condition made part of the sample to understand more accurately the motivations of building renovating profile homeowners. The homeowners' ages were between 29 and 75 years old (12.1% have less than 40; 69.5% were between 40–65 and 18.4% more than 65) and 58.1% were males and 41.9% females. As regarding the households' size, they had from 1–6 members, with 75.5% of them having children living at home. There were homeowners from all educational levels in the sample (60.5% with a secondary or high school level) and 78.1% were employees. The average household income was medium to high level. The houses were, in general, with a regular/good maintenance appearance and have the characteristics of middle-class houses. In addition, 79% of the houses were built or bought and 21% inherited.

5 RESULTS AND DISCUSSION

5.1 An overview about energy-related renovations

Energy-related renovations were considered all the physical changes to the building that involve improvements in the thermal insulation of buildings envelope and in the air and water heating systems. All the other types of renovations were considered the ones that change building aesthetic and functionality that may induce or include energy related improvements. The principal renovation works, frequency and order of priorities adopted by the respondents are in Fig. 1. The aesthetical and indoor renovations are the priority followed by the envelope and outdoors. As far as renewable energies are concerned, the thermal solar collector and the heat recovery systems that close the fireplace are the preferred.

5.2 Social and daily home life practices influencing energy-related renovations

Situations concerning the maintenance, malfunction and failure of buildings components were mentioned in 70.5% of the cases as the trigger to renovation works. Proceeding with an energy renovation normally stems from a renovation conducted out of necessity. One of the examples given by the respondents was the initial need to replace the roof tiles which resulted in the adoption of thermal insulation. Some of the respondents reported a need to solve problems related to condensation in window frames as triggering their replacement. Also, the

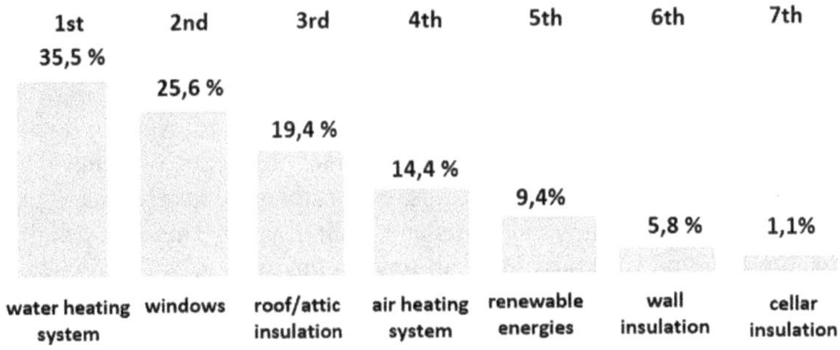

Figure 1: Frequency and preference of energy-related renovations works.

need to replace the air heating system due to failure or in order to achieve higher efficiency made homeowners aware that the next step would be to act on the building envelope thermal insulation in order to reduce heat loss and monetize the new system efficiency.

The analysis conducted showed that Portuguese homeowners start by doing small, sometimes low-cost, renovations out of necessity (not necessarily energetic-related) and gradually continue with other renovations throughout the time (79.3% of the respondents). Some actions encourage and trigger others, depending on the time and money available at that stage, on certain given circumstances of domestic life and on the personal ambition to do certain renovations. In general, it is not only possible to talk about one single event but about some "little-by-little renovation" dynamics, according to the family needs and daily home life actions. From the analysis made to the sequence of measures taken by the respondents throughout the years, small cost-efficient energy efficient measures (even non-physical measures) induced more extensive measures.

Nevertheless, there are some key moments in Portuguese respondents' stage of life that can led to large energy-related renovations. The moment of buying a second-hand house, deciding to live in an inherited house, new relationship, new child, children moving out, the beginning of retirement, widowhood or the beginning/end of a life cycle (emotional stress, family conflicts or disturbance) were important moments detected. Note that 68% of the respondents, who acquired their second-hand home, did some improvement works when moving in, so as to ensure homeowners' comfort and aesthetics requirements. Personal events seem to generate the need to change something in the house. The respondents with prior experiences in some kind of renovation measures showed more determination and less fear to make more energy improvements in the house (64% of them intend to continue renovating). The savings experienced from the adoption of some previous measures has made them more aware of what actions to take next. The lack of experience of some respondents made them acknowledge their reluctance, although they are sensitive to the issue. Most respondents under 50 years old have a shorter timespan for experiences, but as they are in the prime of life, they stated to have plans for the future and aspire to improve their dwelling. Youngers homeowners reveal more enthusiastic.

Another interesting aspect was the homeowners surveyed who stated that they like to plan the improvement of their dwellings, get personally involved in the choice of energy efficient solutions and in the execution of some tasks. They take pleasure from the process as much they do from the result. Pre-acquired knowledge has a paramount influence in these homeowners. They are better informed and are more likely to enjoy DIY tasks. This tendency was observed in a higher scale among the homeowners under 50 years old with a more educational level. They also stated that the construction professionals did not always provide them with the best advice or best solutions. The female respondents, especially those under 50 years old, referred that they like to get involved in the process, but mostly regarding aesthetic issues only.

The homeowners' life stage also seems to be significantly relevant to their attitude. The younger generation of respondents (considered under 50 years old) have grown with a different sensibility and therefore, show higher environmental awareness, more knowledge of energy efficiency and innovative systems and technologies. They have more daily habits connected to this issue and wish to have an environmentally friendly home. On the other hand, the respondents over 50 years old generally stated that what motivates them the most towards renovating is the comfort, safety, easy use (barrier-free) and aesthetics that the building resources provide to their daily life. The physical demands of home life tasks are paramount to the respondents over 65 years old. In this age range, people are more sensitive to temperature variations, they spend more time at home and have more physical limitations. As an example, a number of homeowners stated to have replaced their heating systems with central heating systems, which are more practical and user-friendly (with central controlling devices) and require less physical effort. The switch to more energy efficient equipment using renewable energies or electricity is also a measure frequently mentioned as it they consider less dangerous than gas and more user-friendly for daily use than biomass. The respondents who also gave importance to aesthetics, declare that they are in a position in life where they want to do what they like best and do not mind spending money on it.

Another relevant issue found is that in the renovation project of a residential building there is a family involved, with differentiated individuals and corresponding opinions. The discussion of opinions and the need to reach consensus is acknowledged by the homeowners. Female respondents have more aesthetic and architectural aspirations while male respondents are more concerned and willing to acquire renewable and energy saving technologies. Although the technical solutions do not per se persuade homeowners, the male respondents stated that they might be appealing enough to leverage the renovation process. Most female subjects do not like the untidiness and changes caused by a building renovation in the family daily life (therefore the "little-by-little renovation" appeal), but they are more demanding as far as indoor comfort temperatures are concerned. On the other hand, the male subjects seem more concerned with the investment financial risk.

The data also revealed that the interventions are also motivated by a desire/necessity to maintain or change family life ideals, domestic routines and the way homeowners live their home. Some subjects stated that they intend to improve family socialization practices by creating comfortable, relaxing and informal spaces where the family can gather. Others mentioned a will to host and entertain their families and friends. More natural light and space in the living areas, adopting open spaces like dining room/kitchen, and new rooms for guests, are examples of the home improvements mentioned. In some cases, it was the need to adapt the space to new working and studying practices that led to restructuring indoor spaces. Consequently, such new comfort demands led to pondering ways to improve energy efficiency (envelope thermal insulation and air heating system).

Foreseeing and anticipating changes in home life dynamics was also found to be a factor triggering the wish to renovate, with implications in the use of energy. The need to share spaces was also stated to create conflicts and tensions which can be solved through renovation. Some of the respondents mentioned higher comfort demands due to: household fluctuations (a new child or the return of grown-up children to the family home); school-age children at home or homeowners' retirement. The more time family members spend at home, the higher their energy use. Hence, children and elderly play an important role. A percentage of 72.2% of the respondents under 50 years old stated that increasing their children's level of comfort may be the triggering factor to proceed with home renovations. Nowadays, there are high requirements regarding the temperature of children's bedrooms and living rooms, which implies costs. The elderly, who spend more time at home, also require adequate and comfortable spaces. The respondents under 50 years old also showed concerns with comfortable additional spaces for schoolchildren to play and study. Some respondents (the generation with children up to 18 years old) stated that the current generation of children and teenagers makes an intensive use of their bedrooms, where they have a computer and sometimes even a television set. This means one more room which needs and consumes energy during the day, contrarily to what happened in the past, when most home life activities were held in the kitchen and living room. Generalised comfort in all the household spaces has become a requirement of modern times. This has led homeowners to ponder the adoption of energy efficient central heating systems with the use of cheaper energy or renewable ones.

As far as physical daily tasks are concerned, renovations might oftentimes be triggered by the homeowners' wish for higher comfort in the performance of these tasks. For example, the female respondents claimed for more user-friendly equipment that they need to operate daily. This is the reason given for replacing some appliances with more energy efficient ones (air heating systems, for example) fitted with automatic air temperature control systems. One frequent remodelling action typically conducted in living rooms is to close open fireplaces and install a biomass heat recovery system (e.g. pellets), which is cleaner and more user-friendly. This solution was also considered to be a security measure by the respondents over 50 years old.

One of the most relevant motivations mentioned by the Portuguese respondents concern their ambitions related to the building aesthetics, lifestyle, social status and prestige. The energy-related renovations that brings a better appearance to the house, outdoor and indoor, are very attractive to the homeowners in general (replacement of windows, for example). Undoubtedly, residential buildings are also the expression of the family's identity and oftentimes, dwellers no longer identify with the image that the building transmits. This was another solid reason given by Portuguese respondents for the implementation of physical changes in the buildings, which bring about energy improvements as an additional benefit. Several of the respondents were concerned about the way the home was perceived by neighbourhood. There is a visible trend for the measures connected to energy efficiency to be noticeable to the community which are, frequently, the first ones to be implemented. The adoption of new technologies and renewable energies has been acknowledged as one of most important sign and expression of modernity and environmental awareness (and they are visible). Fashion and trends created by industry (advertising and catalogues) was recognized also as a driver by a great percentage of homeowners. It seems from the analysis of the data that energy efficiency starting to be seen as a trend by the Portuguese respondents and to be considered by homeowner as an instrument to make comparisons with what society defines as an ideal home. One of the favourite renovation actions is the remodelling of kitchens, which are spaces of intense use and also social prestige. Never pay back but provide useful services, convenience and a personal fulfilment. Such renovation sometimes triggers the replacement of household appliances with lower energy consumption devices (for example, the need to the relocate the water heating system may end up leading to its full replacement). The same happens with the renovation of façades (walls and principally windows) and surrounding outer spaces, seen as transmitting something about the building dwellers. Similarly, the renovation of façades led to the application of thermal insulation.

With regard to community life, some respondents admitted that seeing energy efficiency measures being adopted by their relatives, friends and neighbours encouraged them to do the same. The communication within the community network that Portuguese homeowners ends up playing a crucial role in the motivation they have to renovate.

Also noteworthy in the results of this study was the observed will to adopt renovation measures in buildings with higher sentimental value (e.g. inherited buildings). The inherited houses revealed a higher number of non-monetary (without financial return) and aesthetical improvements compared with others cases. As these properties are affectively meaningful to their owners, the set of renovation measures adopted might not even imply any financial return, fact that is irrelevant to the respondents. The choice seems to be emotionally oriented, in this particular case, but these kind of influence were also detected, more or less, in the decisions for all the other types of homes.

Table 1 summarises the influences, together with their relevance, related with homeowner's daily life at home and social practices and norms identified in this preliminary study which reveal to be at the core of decision-making processes regarding the renovation of Portuguese single-family buildings and which had subsequent implications for the adoption of energy-related measures. Some influences stand out: the stage of life and beginning/end of life cycles; family comfort concerns; lifestyles; trends and social standards.

5.3 Practical implications for energy policies

Social sciences have a significant role for informing new policy approaches [12], nevertheless, social analysis is not underlined in the Portuguese public policies about energy

Table 1: Influences, and their importance to the respondents, to renovate single-family buildings with implications in the adoption of energy efficient measures.

	Influencing factors for home renovations with consequences in energy efficient improvements	**% respondents**
Position in life/personal events	Stage of life and the beginning/end of a life cycle	67.4%
	After a personal/family event	42.2%
	Experience in the adoption of previous measures	50.5%
Family dynamics/ social relations	An active participation in the renovation (homeowner's skills)	30.2%
	Roles, relationships and opinions/conflicts within the family	55.5%
	Change/maintain the daily practices and the family life ideals	40.5%
	Concern with the family comfort conditions	82.3%
	Fluctuations in the household composition	55.4%
	Indoor conditions for the family and friends to socialize	30.2%
	Safety and easy-use to manage the home facilities	38.2%
	Building not adapted to dwellers' physical abilities	10.3%
	Higher comfort when performing domestic tasks	40.0%
Lifestyles/ social standards	Improvement of building aesthetics /personal tastes	70.5%
	Achieve social status and new lifestyle	68.3%
	Fashion and trends created by products industry (catalogues)	66.4%
	Building identity not compatible with homeowners' identity	58.3%
	Measures implemented in friends and neighbors' homes	55.2%
	Sentimental meaning and value of the building	52.4%
	Interest of their children to live in their home in the future	28.4%

efficiency for existent buildings. The most recent National Plan for Energy Efficiency [33] reinforces the strategy of mobilize investments to the renovation of the national buildings stock, nonetheless, it keeps on relying on: the usual "save energy" argument used both in the Energy Performance Certificate System (EPCS) and in the informational and awareness campaigns and on financial programmes. All of these tools conceptualized as distinctive and partitioned from other home improvements with which homeowners are engaged throughout the years. If the everyday life at home aspects are involved in dynamic of home improvement, energy-related renovations must be considered in the same way.

The EPC requires that major building renovations be a minimum on efficient classes. The problem is that Portuguese homeowners make primely step by step home improvements, not necessarily in one single action, and where energy efficiency can be included or not. Until now, the contribution for the national energy goals is limited to the implementation of the improvement measures identified on energy certified homes [2]. Nevertheless, Portuguese homeowners still see the EPC as a bureaucratic and complex document, useful (in theory) but in practice with some drawbacks [22], [34]. The EPCs, as all other government informational policy tools (campaigns and informational platforms), lays on the energy saving and profitability arguments to convince the homeowners nonetheless these latter tend to pay more attention to total upfront costs [35] and are sceptical about the costs presented in the EPC [34]. Few Portuguese homeowners require the EPC voluntarily and the majority of the ones that required did not implement the improvement measures recommended [34].

The latest National Building Renovation Plan [2] relies on four specific measures: solar thermal for domestic hot water (DHW); green heat (biomass use in fireplace heat recovery systems); efficient window and thermal insulation. Situating this challenge in a broader understanding, a new conceptualization of these measures can be made. In this study, the Portuguese respondents were willing to perform visible and aesthetical renovations that invoke an image of social status and prestige. The solar thermal for DWH, and other renewable energy systems, in general, and the efficient windows are visible measures, therefore, promoting them in connection with the homeowner's daily life influences linked with lifestyles and social standards could be more accurate strategy. Product industries could be an important partner to awaken the will of the homeowners to the energy renovations creating, as they do for other products, new trends, using appealing images in catalogues with new home solutions, for example. The green heat could be promoted in association with aspects like safety, high comfort when performing domestic tasks and easy-use to manage the home facilities. The roof/attic insulation is not as much as problematic as the wall insulation but both are less visible measures and then it seems not so prioritized by homeowners. These improvements could be incentivized in homeowners that: had implemented previous and related measures (for example, efficient air heating system); have an intense household dynamic that originate high comfort demands and energy consumption (children and older householders) and have family socializing ideals. Finally, but no less important, two imperative issues are: to encourage the start of the step by step renovation process (including energy-related measures or not), even if using an initial small measure, and make all these energy efficiency measures as fashionable as many other areas in homeowners' lives. There are key opportunities for the energy advisor to work with Portuguese homeowners. They could be trained to target homeowners on these specific aspects of their lives because that how it is more likely that energy renovations happen. A special energy advice (door-to-door or available at the municipality) need to reach the homeowners' problems and concerns in order to address the most important aspects for them. Consequently, this calls for more tailored and creative strategies to be developed.

In almost all the European countries, motivational policies are also more focus on techno-economical aspects using the EPCS than in the social and everyday practice context of energy-related renovations [22]. However, there are some countries that are making small steps and leading by the example. Working with the social framework calls for more proximity to the homeowners and better interpersonal communication [13]. Proximity polices and local level actions, with local bodies involved, can play an important role [3]. Local governance strategies taken by some front-runner municipalities try to engage the homeowners with energy efficiency. Personalized information and consultation, group tours, public–private partnerships with local companies; local campaigns and energy advisors prepared with social and communication skills to reach homeowners' needs are examples [3], [13], [22], [24], [30]. Local offices for free energy advice and consultancy and local workshops/exhibitions with the participation of opinion leaders is a work in progress in Sweden [13]. In Denmark, where EPCs have the longest implementation, promote energy-related renovations by offering free energy advice at home is a strategy already in the field, studied and monitored on a social and everyday life perspective [15].

6 CONCLUSIONS

As we saw previously, the energy-related renovation seems to be strongly related with everyday life of Portuguese homeowners. Houses are the reflect of homeowner's life stage, family dynamics, social relationships, lifestyles and social rules. This paper aimed to bring social science to the discussion in order to get an in-depth knowledge about how Portuguese homeowners' home life conditions and everyday social practices influence energy-related renovations in single-family buildings. This study contradicts the assumption of this is an investment that is guided fundamentally by financial restrictions. The homeowners regard several daily life and non-profitable factors when they decide to renovate which can increase the motivation to implement energy efficient measures. Renovations for reducing Portuguese residential buildings' energy consumption must be seen by policy-makers as part of a continuous process of gradual adaptation of the building to family life. The most important highlights can be recorded. Energy efficiency renovation cannot be seen as one single isolated event, but as a "little-by-little" practice conducted jointly with or induced by other remodelling approaches involving multiple aspects of the homeowner's daily life negotiated within the family. Encourage the start of this continuous and gradual process of renovation is highly important. One initial measure, even simple, can activate many others in the future. The results also demonstrate that is great the influence that aesthetic, trends and social standards have in homeowners' decision. Make energy efficiency a fashionable issue could be an interesting challenge for Portuguese policy-makers, without forgetting to work with the industries that have huge influence and know how to shape people tastes and tendencies. Something saw in a magazine or catalogue can be a powerful tool. Information and communication policies must be structured and adapted to the different social context of the homeowners and is expectable that local bodies can have an important role on that.

REFERENCES

[1] Eurostat, Distribution of population by tenure status, type of household and income group: EU-SILC survey. http://appsso.eurostat.ec.europa.eu/nui/show.do?dataset=ilc_lvho02&lang=en. Accessed on: 2 Nov. 2018.

[2] Direção Geral de Energia e Geologia, *Estratégia Nacional para a Renovação de Edifícios*, Direção Geral de Energia e Geologia: Lisboa, 2014.

[3] Stieß, I. & Dunkelberg, E., Objectives, barriers and occasions for energy efficient refurbishment by private homeowners. *Journal of Cleaner Production*, **48**, pp. 250–259, 2013. DOI: 10.1016/j.jclepro.2012.09.041.
[4] Gillingham, K. & Palmer K., Bridging the energy efficiency gap: policy insights from economic theory and empirical evidence. *Review of Environmental Economics and Policy*, **8**(1), pp. 18–38, 2014. DOI: 10.1093/reep/ret021.
[5] Whitmarsh, L. et al., *Public Attitudes, Understanding, and Engagement in Relation to Low-Carbon Energy: A Selective Review of Academic and Non-Academic Literatures*. Research Councils UK (RCUK): London, 2011.
[6] Moezzi, M. & Janda, K.J., From "if only" to "social potential" in schemes to reduce building energy use. *Energy Research and Social Science*, **1**, pp. 30–40, 2014. DOI: 10.1016/j.erss.2014.03.014.
[7] Wilson, C., Crane, L. & Chryssochoidis G., Why do homeowners renovate energy efficiently? Contrasting perspectives and implications for policy. *Energy Research and Social Science*, **7**, pp. 12–22, 2015. DOI: 10.1016/j.erss.2015.03.002.
[8] Maller, C.J. & Horne, R.E., Living lightly: How does climate change feature in residential home improvements and what are the implications for policy? *Energy Research and Social Science*, **1**, pp. 41–48, 2014. DOI: 10.1080/08111146.2011.539514.
[9] Gram-Hanssen, K., Retrofitting owner-occupied housing: remember the people. *Building Research and Information*, **42**, pp. 393–397, 2014. DOI: 10.1080/09613218.2014.911572.
[10] Judson, E.P. & Maller, C., Housing renovations and energy efficiency: Insights from homeowners' practices. *Building Research and Information*, **42**(4), pp. 501–511, 2014. DOI: 10.1080/09613218.2014.894808.
[11] Lutzenhiser, L., Through the energy efficiency looking glass. *Energy Research and Social Science*, **1**, pp. 141–151, 2014. DOI: 10.1016/j.erss.2014.03.011.
[12] Gram-Hanssen, K. & Georg, S., Energy performance gaps: promises, people, practices. *Building Research and Information*, **46**(1), pp. 1–9, 2018. DOI: 10.1080/09613218.2017.1356127.
[13] Nair, G., Gustavsson, L. & Mahapatra, K., Factors influencing energy efficiency investments in existing Swedish residential buildings. *Energy Policy*, **38**(6), pp. 2956–2963, 2010. DOI: 10.1016/j.enpol.2010.01.033.
[14] Stern, P.C., Individual and household interactions with energy systems: toward integrated understanding. *Energy Research and Social Science*, **1**, pp. 41–48, 2014. DOI: 10.1016/j.erss.2014.03.003.
[15] Tjørring, L. & Gausset, Q., Drivers for retrofit: a sociocultural approach to houses and inhabitants. *Building Research and Information,* **47**(4), pp. 394–403, 2018. DOI: 10.1080/09613218.2018.142372.2
[16] Walker, G., The dynamics of energy demand: change, rhythm and synchronicity. *Energy Research and Social Science*, **1**, pp. 49–55, 2014. DOI: 10.1016/j.erss.2014.03.012.
[17] Gram-Hanssen, K., Understanding change and continuity in residential energy consumption. *Journal of Consumer Culture*, **11**(1), pp. 61–78, 2011. DOI: 10.1177/1469540510391725.
[18] Shove, E., *Comfort, Cleanliness, and Convenience: The Social Organisation of Normality,* Berg: Oxford, 2003.
[19] Stirling, A., Transforming power: Social science and the politics of energy choices. *Energy Research and Social Science*, **1**, pp. 83–95, 2014.

DOI: 10.1016/j.erss.2014.02.001.
[20] Guy, S. & Shove, E., *The Sociology of Energy, Buildings and the Environment: Constructing Knowledge, Designing Practice,* Routledge: London, 2000.
[21] Karvonen, A., Towards systemic domestic retrofit: a social practice approach. *Building Research and Information*, **41**(5), pp. 563–574, 2013. DOI: 10.1080/09613218.2013.805298
[22] Bartiaux, F., Gram-Hanssen, K., Fonseca, P., Ozoliņa, L. & Christensen, T.H., A practice-theory approach homeowners' energy retrofits in four European areas. *Building Research and Information,* **42**(4), pp. 525–538, 2014. DOI: 10.1080/09613218.2014.90025.
[23] Vlasova, L. & Gram-Hanssen, K., Incorporating inhabitants' everyday practices into domestic retrofits. *Building Research and Information*, **42**(4), pp. 512–524, 2014. DOI: 10.1080/09613218.2014.907682.
[24] Gram-Hanssen, K., Existing buildings: Users, renovations and energy policy. *Renewable Energy*, **61**, pp. 136–140, 2014. DOI: 10.1016/j.renene.2013.05.004.
[25] Shove, E. & Pantzar, M., Consumers, producers and practices: Understanding the invention and reinvention of Nordic walking. *Journal of Consumer Culture*, **5**(1), pp. 43–64, 2005. DOI: 10.1177/1469540505049846.
[26] Strengers, Y., Nicholls, L. & Maller, C., Curious energy consumers: humans and nonhumans in assemblages of household practice. *Journal of Consumer Culture*, **16**(3), pp. 761–780, 2016. DOI: 10.1177/1469540514536194.
[27] Maréchal, K. & Holzemerb, L., Unravelling the 'ingredients' of energy consumption: Exploring home-related practices in Belgium. *Energy Research and Social Science*, **39**, pp. 19–28, 2018. DOI: 10.1016/j.erss.2017.10.025.
[28] Palm, J., Ellegård, K. & Hellgren, M., A cluster analysis of energy-consuming activities in everyday life. *Building Research and Information,* **46**(1), pp. 99–113, 2018. DOI: 10.1080/09613218.2017.1302225.
[29] Wilson, C., Pettifor, H. & Chryssochoidis, G., Quantitative modelling of why and how homeowners decide to renovate energy efficiently. *Applied Energy*, **212**, pp. 1333–1344, 2018. DOI: 10.1016/j.apenergy.2017.11.099.
[30] Galvin, R. & Sunikka-Blank, M., The UK homeowner-retrofitter as an innovator in a socio-technical system. *Energy Policy*, **74**, pp. 655–662, 2014. DOI: 10.1016/j.enpol.2014.08.013.
[31] Bravo, G., Pardalis, G., Mahapatra, K. & Mainali, B., Physical vs. aesthetic renovations: Learning from Swedish house owners. *Buildings 2019,* **9**(1), p. 12, 2018. DOI: 10.3390/buildings9010012.
[32] Janson, U., Passive houses in Sweden: Experiences from design and construction phase. Report EBD-T--08/9, Lund University, 2008. https://docplayer.net/22330807-Passive-houses-in-sweden.html. Accessed on: 23 Feb. 2019.
[33] Plano Nacional de Ação para a Eficiência Energética (PNAEE). https://ec.europa.eu/energy/sites/ener/files/documents/pt_neeap_2017_pt.pdf. Accessed on: 6 Apr. 2019.
[34] ADENE, Estudo de Opinião no Âmbito de uma Campanha de Sensibilização e Promoção da Eficiência Energética na Habitação Particular, ADENE. www.adene.pt/edificios/. Accessed on: 22 Feb. 2019.
[35] Baginski, J.P. & Christoph, W., A Consumer decision-making process? Unfolding energy efficiency decisions of German owner-occupiers. HEMF Working Paper, No 08/2017, 2017. DOI: 10.2139/ssrn.3023997. Accessed on: 15 Feb. 2019.

LESSONS FROM DEPLOYING LARGE-SCALE SOLAR ELECTRIFICATION IN BANGLADESH: CAN THE LAST MILE BECOME THE FIRST?

GEORG HEINEMANN, RALUCA DUMITRESCU, CHRISTIAN VON HIRSCHHAUSEN, NOARA KEBIR & DANIEL PHILIPP
TU Berlin, Microenergy Systems Research Group, Germany

ABSTRACT

With over four million solar home systems (SHSs) deployed, Bangladesh stands out in the development of decentral, bottom-up electrification. Amid a difficult socio-economic and political environment, joint efforts by local authorities, financing institutions, entrepreneurs and engineers, supported by international donors, a community-based infrastructure development, largely based on local value creation, has succeeded. This paper discusses the success factors of the Bangladeshi SHS development over the last two decades and its current ambitions, focussing on the organizational models deployed, amongst others, on value-added production and financing. We then provide a critical assessment of the perspectives to make the last mile the first one, i.e. to use SHS as the driver of decentral, yet grid-based electrification. Last but not least, we give an outlook towards the potential of "swarm electrification", i.e. the bottom-up (r)evolution of the energy infrastructure by interconnecting existing and new electricity usages, storages and generation, enabling a peer economy based on prosumerism and local value creation.

Keywords: bottom-up electrification, off-grid organizational model, IDCOL SHS program, decentral infrastructure development, microgrid, swarm electrification, last-mile distribution.

1 INTRODUCTION

Since its independence from Pakistan in 1971, the country of Bangladesh has been challenged by its small size, its limited resources and a population density, which makes it one of the most populous countries in the world. Although its socio-economic situation has substantially been improving over the last years, Bangladesh is still ranked 139 out of 188 countries on the human development index. In rural Bangladesh, where more than 60% of the population lives, it was in principal too laborious and expensive to extend the national electricity grid [1]. As a result, 42.49% of them had still no access to electricity in 2010 [2]. Six years later, the access rate has almost doubled and an access to electricity rate of 69% in rural Bangladesh can be noted [3]. The key pillar of this achievement is the large-scale and decentral deployment of solar home systems (SHSs). Already back in 2003, Bangladeshi authorities initiated a publicly co-funded SHS loan program, managed by the public Infrastructure Development Company Limited (IDCOL). The IDCOL SHS program is characterized by local stakeholders implementing the program in a decentralized manner and by a highly standardized SHS, which is packaged with a simple consumer loan for the end-users. Over the period of almost two decades this setup, co-funded by various international development partners, has gained considerable momentum, and by 2017, more than 4 million SHSs had been installed [4]. Bangladesh has been recognized as the largest solar off-grid program as well as the country with the highest share (9%) of total population connected to off-grid supply in the world [5].

2 THEORETICAL BACKGROUND

In changing and complex environments, the provision of public infrastructure such as rural electrification faces a number of challenges, such as high investment costs, continuous and

WIT Transactions on Ecology and the Environment, Vol 237, © 2019 WIT Press
www.witpress.com, ISSN 1743-3541 (on-line)
doi:10.2495/ESUS190071

costly maintenance and long amortising periods. Furthermore, the inherent complex nature of providing electrification on household level by using SHS implies additional high transaction costs [6]. Identifying the different elements and their characteristics as well as understanding motivation and behaviour of the involved stakeholders is the principal objective of our investigation. Thus, the starting point of our analysis is the understanding of the "system" of the public good – all the elements involved to provide rural electricity by deploying solar systems on household level. Its complexity requires distinguishing between the various parts of this task. Beckers et al. [7] define the provision of a public good as a transformation process with different phases. Input has to be identified, pooled and interrelated to achieve the desired output. Key elements are the technical system, actors, their assets and their roles, as well as institutions.

The technical system describes the targeted public good in combination with its characteristics such as assets and basic processes to produce the good. Actors can be individuals or organizations. They possess their own set of objectives, for example maximizing profit, as well as specific assets and capabilities which contribute to the transformation process. Actors assume roles. Roles represent packages of sub-tasks within the transformation process. Roles can be organized according to their package of tasks or according to the respective transformation phase they fit into. The nature of actors and their capabilities as well as the complexity of the technical system create interdependencies. Therefore, institutions are needed to assume the role of coordination between them. Institutions are formal or informal norms which frame the interaction between the elements, the actors and their roles.

The second dimension is the discussion of the organizational model. The provision of infrastructure such as rural electricity can be organized by the government, the private sector or by choosing a hybrid model from these two options. While researchers form a common understanding on the complexity of choosing a model for making infrastructure available, findings such as the research from Hassheider [8] on the provision of transregional road infrastructure, Ewers [9] on privatizing federal highways and Ménard and Saleth [10], which discuss alternatives for the provision of potable water, are not able to recommend a universal organizational model for public infrastructure tasks. Finding the most appropriate organizational model strongly depends on the actual characteristics of the project and the local conditions [11]. The theory of new institutional economics (NIE) is to lead to an understanding of how the strategy of institutional organization for decentralized electrification might look like. The objective of NIE is to investigate institutions and their interaction with organizational structures. Institutions describe constructs of formal and informal norms and rules, as well as the organization of their implementation within a society. This reflects the advancement of the theories of neoclassical economics. NIE not only looks at the pure rational approach of neoclassical theory, but also incorporates actual realities such as time, individual and emotional behaviour, as well as other transaction costs. Numerous scientific works already deal with new institutional economics as the underlying theory for providing public infrastructure [6], [10]–[12].

Furthermore, to classify organizational models Klatt [13] and Wealer et al. [14] distinguish between the categories provision, in which the actual need is defined, production, consisting of the transformation process of the public good, and financing. The transformation process can be further characterized as an entire value chain consisting of the five phases planning, production, distribution, operation and maintenance. The interdependencies and the structuring of these three categories is supposed to shed light on the principal question of how the decision-making process can be designed to achieve the intended output.

3 METHOD

Our methodology is based on a research design which reviews existing literature on rural electrification programs as well as the theory of organizational models in public infrastructure. We apply a descriptive-longitudinal case study approach by reviewing the different stages of rural electrification in Bangladesh. The underlying objective is to discuss the development of the SHSs program in Bangladesh and its related organizational model. Our principal unit of analysis is the historic and contextual component. Among others, our analysis includes extensive findings from the World Bank [15]–[17] as well as dedicated policy research papers [1], [18]–[20] to explore the evolution of the program since its inception in 2002. Special attention is paid to involved stakeholders. While literature on business models of SHS programs is vast, the institutional perspective has received much less attention. Thus, as a second step, we explore the program according to the organizational model deployed. Lastly, we introduce swarm electrification as one solution to the recent challenges of the program. Bottom-up grid infrastructure by interconnecting solar home systems may present the next level of sustainable off-grid electrification. Our methodological concerns have consequently followed a two-stage approach. A consistent overview of the most relevant literature in Bangladesh as well as a discussion of organizational models in the field of rural electrification was reached.

4 FINDINGS

By analysing SHS electrification in Bangladesh over the course two decades, our study sheds light on the organizational model deployed. More specifically, our findings reveal the three critical dimensions of organizing the process of SHS deployment: (i) the production, (ii) the provision and (iii) the financing. In the sections that follow we introduce the IDCOL SHS program, map the underlying organizational model and discuss the historic development of the three dimensions in turn. Our historic analysis will be categorized according to (i) early days (2003–2008), (ii) take-off and peak (2009–2015) and (iii) recent developments (2016–present).

4.1 The IDCOL SHS program

The IDCOL SHS program started in Bangladesh in 2003. Since then it has been implemented by the state-owned IDCOL. Its capacity originates in its public mandate of financing infrastructure projects and in its right of extending soft loans to private companies. As a non-banking institution ICDOL is supplied with funds by the Government on the one hand, and by international donor organisations, on the other hand. For the SHS program, IDCOL uses these funds to provide loans at non-market conditions to the Partner Organisations (POs). The POs use these funds to pre-finance the required SHS provided by the suppliers and to set up affordable consumer loans for the targeted off-grid households. The POs are the single point of contact for the households. They carry out installation, user training, maintenance and loan collection.

The program's underlying organizational model is organized according to two premises (Fig. 1): first, the elements to provide the output, rural electrification with SHSs on household level, are organized horizontally along the defined value chain. The value chain not only divides the transformation process in five phases, but also classifies the actors according to their roles during each phase. Interdependencies exist on the level between IDCOL/government and the organizations (suppliers and POs), between POs and suppliers, as well as between POs and households. Second, our three principal "organizational" dimensions are mapped, too: the financing perspective is fully organized and provided by

IDCOL. The POs are in charge of the provision, which includes the distribution, operation and maintenance phases. They are selected and mandated by the IDCOL's PO Selection Committee. Lastly, private manufacturers are in charge of the production. IDCOL's Technical Standards Committee provides directives and approvals for them [18]–[20].

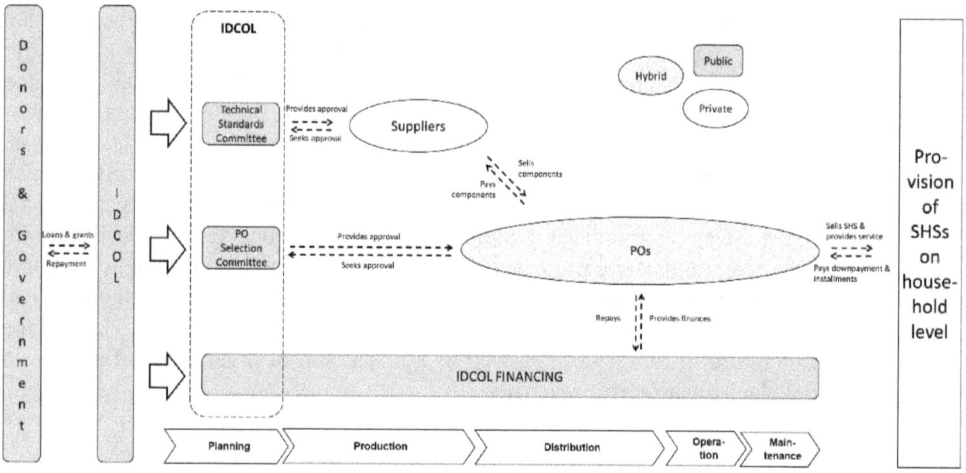

Figure 1: Organizational model of the Bangladesh SHS program. *(Source: Own compilation.)*

4.2 Production

The production perspective is principally influenced by three actors: IDCOL, local suppliers and global suppliers.

4.2.1 Phase 1: early days

A dedicated local industry focusing on the manufacturing and assembling of solar home systems did not exist when the program started in 2003. In 2002, the World Bank conducted a survey and concluded that only a very negligible number of battery manufacturers and fluorescent lamp producers were existent [16]. Thus, in its beginning, the program relied mostly on importing SHS components from global manufacturers based in countries like Singapore, India and China [19]. In the following IDCOL established a dedicated Technical Standards Committee (TSC). First of all, TSC was assigned with the task to manage the supply side by approving suppliers and ensuring technical standards. Procedures, inspectors and technical experts ensured minimum quality of (local) equipment and after-sales services incl. maintenance and product warranties for a minimum of 3 years. Random technical audits, a dedicated customer call center at IDCOL head office and 13 regional offices were further examples for TSC's role [17]. However, IDCOL and its executive arm TSC as its executive arm TSC started various support initiatives, too, which actually promoted the local solar industry in the end. The renewable energy authority SREDA was initiated, tax reduction and special incentives for local manufacturing of renewable products and components were introduced, private sector investment was facilitated, a national target plan for renewable energies was launched and import tariffs for assembling solar components were reduced.

4.2.2 Phase 2: take-off and peak

While in the first five years of the program a total of 200,000 SHSs were installed, the real take-off started in the period 2008–2011. In just 3 years a total of 1.4 million SHSs were deployed [19]. IDCOL's policies to foster local manufacturing, which reduced end-user prices as well as improved the efficiency of the supply chain plus a substantial global decrease of solar component costs contributed to the boom.

An example for IDCOL's effective framework was the mandatory local sourcing and assembling of charge controllers. It turned out that local quality was very low at the beginning. The return rate was so high that each PO had to compulsively hold a decentralized emergency stock of 2% of their turnover available. The replacement to end-users had to be provided free at all times. Later, the emergency stock was reduced to 1%. It reflected a powerful intervention to improve quality of the locally made charge controllers resulting in 53 local suppliers at the peak of the program. Various aiding measurements grew out of it, such as mandatory free emergency stock for other components, engraving customer service hotline numbers instead of stickering, as well as an overall decrease of the mean time to failure rate. They confirm the evolvement of the local solar industry during that time [1].

Another example is the local manufacturing of the battery component. Batteries are the most delicate and vulnerable part of the SHS, as they can break down quickly due to poor battery quality and management. And they may represent up to 40% of the total system cost. Managing this component well is one of the key success factors of the IDCOL program. The type of battery to be provided was stipulated by IDCOL. POs were obliged to use tubular plate lead-acid batteries. They have the benefit of allowing regular discharge of 80%, featuring a higher charging efficiency and a low rate of self-discharge. Batteries were to be provided with a 5-year warranty to the end-user. As they were locally available, IDCOL decided to only approve local manufactured batteries for the program. When the program started, only 3 local suppliers were approved by IDCOL. At the peak of the program a total of 30–40 suppliers existed, 22 of them were ISO certified. Recycling schemes were enforced, too. In 2013 three recycling plants were existent highlighting IDCOL's unique battery replacement policy. POs were urged to notify SHS users 3 months prior to the estimated ceasing of the battery. They would then receive a 24% rebate on a new battery if they returned the old one. IDCOL triggered this proactive after-sales program by providing POs and battery manufacturers a USD 10 supplement for each newly installed battery plus a 100% refinancing option [19], [20].

The battery component was one of the trigger points for the take-off of the program establishing and confirming an efficient framework of quality control and customer service. Being so critical for the performance of the system, IDCOL's proactive battery management was essential for achieving positive end-user perception and in a rather short period of time rural Bangladeshis learned to trust and appreciate IDCOL's SHSs.

4.2.3 Phase 3: recent developments

By 2015, 31,750 jobs existed in the solar manufacturing sector. Next to 30–40 battery suppliers, the local industry comprised of 57 LED suppliers, 53 charging controller suppliers and 9 solar panel assembly plants. While batteries, LED suppliers and charging controllers were 100% locally sourced, the local solar panel assembling covered approximately 20% of the total PV demand [19]. Nevertheless, sales of SHSs are recently slowing down and the number of jobs are slightly decreasing, too [21]–[23]. Further details and reasons will be discussed in the following section.

To sum up, the IDCOL program can be overall regarded as a strong industrial policy. Local manufacturing and jobs were created. Domestic research and development were fostered too. A considerable cost reduction of components was achieved. SHSs in Bangladesh cost nowadays only about 50% of SHSs available in East Africa [24]. Furthermore, the growth of the local supply side helped to improve and adjust the technology to the local needs and to introduce valuable accessories. Analysing the production perspective illustrates that implementing and fostering local manufacturing requires economic support and protectionism on local level combined with a certain time horizon.

4.3 Provision

Provision describes, as we have defined in the beginning, the dimension to distribute the public good in question. Looking at our organizational model, deploying SHSs in rural Bangladesh is mainly facilitated by the POs. POs are IDCOL's partner organizations that execute the distribution and installation of the SHSs on household level. POs can be different type of organizations such as for-profit SMEs, social enterprises, NGOs and MFIs. One thing they all do need to have in common: a very rural, a decentralized distribution model, with a strong presence in the countryside. Furthermore, they are regulated by the Microcredit Regulatory Authority of Bangladesh and they are guided by a well-defined Memorandums of Understanding (MoU) with IDCOL [20].

In Bangladesh, the last-mile provision of the SHSs by the POs is one of the key drivers of success.

4.3.1 Phase 1: early days

Grameen Shakti and BRAC were the only distributors which started operations before the IDCOL SHS program was initiated in 2003. A closer look at Grameen Shakti (GS) helps to illustrate the distributor perspective in the early days. GS is a 1996 spin-off of the well-established Grameen Bank. GS, which means in Bengali "village energy", started off with the idea to migrate the successful MFI model of Grameen Bank into the energy sector. Its business model was based on the lessons-learnt of the Grameen Bank summarized in the following three principles [25]:

1. The SHS user pays a down-payment and commits to monthly instalments.
2. GS staff provides service, incl. installation, maintenance, credit collection
3. The SHS is fully owned by the user. The fee-for-service model, which was favoured by the government at that time, was not considered.

Despite of this initial strategic alignment with their experience from the MFI sector, GS discovered key barriers to rapidly increase SHS installations. First, GS faced the challenge of providing appropriate technology for their consumer loans. Solar technology was still in its infancy in the nineties. Providing cost effective and reliable solar home systems was a major challenge to GS. Most of the components had to be imported at high cost and had to be packaged/customized as needed, although local know-how was not yet available. Second, GS struggled with low awareness for solar-technology and solving this was hampered by a substantial lack of marketing and of capacity building abilities [16]. Third, GS was not able to offer attractive conditions to its target users. GS started with a simple SHS package consisting of CFL bulbs, battery, charging controller and solar panel. The package was offered at around USD 500. Considering the limited rural income levels, GS offered the SHS on a consumer loan, but only with loan tenure of max. 12 months. These conditions depended on GS's own refinancing situation with its mother entity, the Grameen Bank. The typical MFI business model extends 1 year business loans at USD 100–200 at a 20% interest for starting or refinancing micro-businesses [25]. GS could not extend very different conditions to its customers. Fourth, cost of operating a distributed SHS business model was extremely high. GS was only selling about 20 SHSs per month. The business model was rather based on low margins considering high technology cost vs. low residential purchasing power. GS could not reach economies of scales as hoped. Moreover, the complex value chain in terms of sourcing the systems, installing and delivering them to remote place as well as frequent after-sales and maintenance visits, turned the GS business case into an unappealing and unprofitable business.

4.3.2 Phase 2: take-off and peak

One of the major obstacles for electrification through SHSs is the establishment of a robust supply chain at an affordable end-user price. The IDCOL SHS program solved this equation by relying on the pertinent pre-program experiences of GS. IDCOL introduced an attractive refinancing scheme, which enables the POs to organize the last-mile-distribution on large scale, offering technical services (product, installation, after-sales services) and financial services (loan provision, instalment collection) to rural Bangladeshis. The POs not only benefit from increased sales but also from economies of scale as well as direct financial support for capacity building.

Competition played an important role to kick-off the program, too. The program started with only five players to test and fine-tune the model. As soon as it proved itself to be working, IDCOL introduced the strategy of excess of supply. Low entry barriers combined with start-up incentives for new distributors proved to be effective. IDCOL aspired to have as many POs as possible. The demand to be satisfied was enormous and the challenge to reach it with an efficient last-mile distribution was complex. In addition, by promoting strong competition, IDCOL was hoping to introduce an additional, self-embedded shakeout mechanism. A broad and competitive supply would quickly eliminate players offering sub-standard quality or overpriced services [20]. Asaduzzaman et al. [18] discuss the usefulness of this strategy. On the hand, it enhances quality and service. It makes sure the subsidy really reaches the end-user. They argue that GS was operating almost alone at the beginning. From 2009 onwards, more and more players joined distribution. These new entrants received more favourable financial conditions. They were able to offer more competitive products and services. Respectively, during this phase end-user prices decreased and SHS installations reached a total of 3.5 million as of December 2014. At the same time the number of active POs reached 56 and altogether they disposed of a very rural and extensive sales and after-sales network with up to 5,732 sites as of September 2015 [19].

4.3.3 Phase 3: recent developments

From 2013–2015 annual SHS installations have sharply decreased by 48%. Ever since then, the installations rates have not recovered and POs are substantially affected. Reactions are mixed. Most of the POs see the entry of cheap generic products as the main root of the problem. Their bad quality is hampering the reputation of SHSs in the market. Furthermore, they are convinced that it is IDCOL responsibility to solve this issue by enforcing more restrictive quality and pricing conditions, especially on the ignored but growing generic market. Only few POs are trying to respond proactively to the challenge. An example is Rahimafrooz, the subsidiary of the biggest battery supplier in the market. Next to GS, they became the leading PO over the years. They are tackling the crisis by trying to put more emphasis on innovation. They just started to digitize the payment data of their clients. They have also introduced a sales app for its agent network and equipped all agents with smartphones. These digitization efforts are supposed to take out inefficiencies in the value chain, but they also hope to strengthen the relationship with their customers. They target to explore it for new and smart avenues of income such as using the digitized customer credit data for introducing banking or furniture products [26].

4.4 Financing

As already mentioned, when discussing the success of the POs, the financing perspective is the most critical component of the IDCOL SHS program. Effective aggregation, management and provision of funds are key drivers for successful SHS programs [15], [27]. Our historic analysis of the financing perspective focuses on the costing and raising of the required funds, embedding financial instruments into the value chain to provide access to finance wherever required, as well as on the management and control of the complete financing scheme.

4.4.1 Phase 1: early days

At the beginning, IDCOL joined forces with the strategic development partners German Development Cooperation (GIZ), Global Environment Facility, International Development Association, and Kreditanstalt für Wiederaufbau (KfW) to set-up a feasible financing scheme [19]. The collaboration addressed the following key components:

Management structure: Although receiving funds and technical support from the development partners, the overall management and control of the financing scheme remained in the hands of IDCOL. This approach followed the concept of external segregated fund management (Wealer et al. [14]) resulting in a centralized funds scheme managed by IDCOL with the benefits of increased fund transparency, centralized management avoiding decentral mismanagement or even bankruptcy and improved public confidence in the program.

Calculating the cost and scope of the program: Furthermore, Wealer et al. [14] defines five approaches for estimating the costing of infrastructure. The IDCOL financing approach was a blend of the specific analogy approach and the bottom-up technique. The bottom-up technique looks at the smallest components of the cost. The specific analogy compares the costing to similar projects in the past. IDCOL could rely on the experience of GS in the 1990s as well as the government projects in 1997, when a SHS pilot for 850 households in Narsingidi District was conducted, and in 2002, when the program "Diffusion of RE technologies" for 6,000 households was launched [19]. Combining both approaches IDCOL came up with a scheme to implement a microfinance-based, direct sales program. The initial scope targeted to deploy 50,000 SHS by 2008.

Timing: The new financing scheme incorporated a soft loan facility provided by the various international development partners. Thus, IDCOL and its implementing POs were able to change the terms of its SHS consumer loans. User's repayment period was extended to three years. Consequently, POs could finance their working capital through IDCOL over a period of up to 10 years. Not only enabled this extended loan tenure households to afford the monthly instalments but also it facilitated sufficient and sustainable refinancing for the production and provision side [20].

In August 2005, the initial milestone of 50,000 SHS was already achieved, 3 years earlier than planned and USD 2 million below estimated costs [19].

4.4.2 Phase 2: take-off and peak

During the take-off and peak period, the specific IDCOL strategy, providing a mix of output-based subsidy, supported refinancing scheme, and institutional and technical incentives, proved to be highly efficient. Rapid take-off and solid market creation was facilitated by offering three principal financial instruments [20]:

1. Buy-down grant: To subsidize rural electricity access on household level, IDCOL started to provide a fixed capital buy-down grant at USD 70 per system. It was an upfront grant, which went directly to the PO in the moment the installation was confirmed.
2. Refinancing: In 2003, POs were able to refinance 80% of the SHS retail price with a subsidized loan provided by IDCOL. The soft conditions were a reduced interest rate of 6% and extended loan tenure of up to 10 years. The remaining 20% of the SHS retail price had to be covered by the customer's down payment and/or be refinanced by the PO itself. This "market" component ensured that ownership for the system and the extended loan among users and POs were maintained. Furthermore, POs were able to increase their profitability by leveraging their IDCOL refinancing loan against the consumer loan. The SHS consumer loan had a higher interest rate and a shorter lending period. The resulting spread could be used to increase the profit margin.
3. Institutional grant: An institutional grant of USD 20 for each SHS installed for the POs was available in 2003. It further subsidized the affordability of the SHS, but most of all it represented a direct support for the PO to build its operational capacity over time training of employees, improving company's assets/operational infrastructure (IT, vehicle fleet) are some examples. But this subsidy was only disbursed on a cost sharing basis of 80% (IDCOL)/ 20% (POs) on the one hand, and if the POs non-performing loan portfolio was low enough, on the other hand. In this way, IDCOL managed to include a self-embedded control mechanism to monitor the performance of its distribution partners.

Asaduzzaman et al. [18] discuss the effectiveness of the buy-down grant. The grant per SHS was fixed, favoring the price of smaller systems resulting in more affordable offerings for poorer households. However, working independently and with a direct-sales program, POs certainly did try to increase prices whenever possible. Therefore, the grant could have had the exact opposite effect in the end. The analysis concludes that competition is required to avoid the described effect, which confirms the important aspect of creating a competitive environment among POs as we have already described in Section 4.3.2.

In addition to this analysis, a closer look at the development of the financing provided by IDCOL to the POs illustrates the effectiveness of this subsidy scheme (Tiedemann [20]): With the above described conditions for PO refinancing (loan) and grant (USD/system installed), total annual installations of 11,697 SHS were reached in 2003. Already in 2004 both grant components were reduced for the first time and were gradually phased out in the following years. In 2013, while refinancing for the POs remained at similar conditions, the grant component was almost non-existent, only a fixed USD 20 subsidy for systems <30Wp remained. However, the annual installations reached an all-time high of a total of 852,388 SHS. Although Tiedemann's [20] calculation is only an approximation, it illustrates the importance of the refinancing loan. In the moment the maturity of the loan was largely extended SHSs installations increased drastically over the course of the years. Reducing the fixed grant subsidy per system installed from USD 90 in 2003 to almost zero on the contrary, did not have any negative effect at all.

4.4.3 Phase 3: recent developments

Over the years, the program has substantially increased its scope. It is nowadays supported by a variety of development partners including the Asian Development Bank (ADB), Department for International Development of the United Kingdom, GIZ, Global

Environment Facility, Global Partnership on Output-Based Aid (a multidonor trust fund), Islamic Development Bank, Japan International Cooperation Agency, KfW, SNV, United States Agency for International Development, and World Bank. The financing scheme was extended several times. Its total targeted investment of USD 191 million at the beginning has reached more than USD 700 million by now providing over 4 million households with electricity access. Reflecting an estimated cost to government of USD 170/SHS installed, the IDCOL program is not only one of the largest but also one of the most cost-effective programs in the world [4], [26].

Drawing on this historic overview, our analysis demonstrates that the IDCOL financing scheme has been pivotal for the scale and outreach of SHS installations in Bangladesh. By centrally managing the funds, adapting loan conditions on PO and household level to the actual needs, the program has been successful in setting up a SHS market for more than four million users.

5 OUTLOOK

On 21st of November 2017, the United Nations Department of Economic and Social Affairs awarded two companies in Bangladesh the "Powering the Future We Want" prize for their joint project that allows villagers to secure more solar power, save money through flexible energy access, and generate income from trading electricity [28]. The two companies, a solar technology developer and the largest PO, Grameen Shakti, interconnect existing SHS into peer-to-peer (P2P) swarm microgrids, allowing end-users to trade the self-generated surplus electricity among each other, in a process called Swarm Electrification. Swarm Electrification is the bottom-up evolution of the energy infrastructure by interconnecting individual existing and new electricity usages, storages and generations, enabling a local shared economy based on prosumerism and local value creation. Through these P2P swarm microgrids, network effects are generated through the inclusion of local economies with strong producer-consumer linkages embedded within larger systems of trade and exchange.

Although the P2P energy concept is more often encountered in the Global North, due to its particular energy infrastructure developed through the IDCOL SHS program, Bangladesh is the first country in which the initiative was implemented. The pre-existence and the density of the standardized and quality certified standalone energy systems constitute the technical infrastructure as the basis for Swarm Electrification. However, within this vast network of SHS, the households do not fully utilize the electricity stored in their battery, resulting in a full battery by midday, limiting the generation potential of their systems. At the same time, other households and businesses require electricity beyond what their systems can supply but cannot afford the acquisition of a new solar system. Through the P2P swarm microgrids, households and businesses are interconnected bottom-up, granting for the trading of the locally generated electricity through blockchain based applications. Such grids grow incrementally towards the national grid, allowing communities that would otherwise represent the last mile of rural electrification to become the first mile. The economic infrastructure of Swarm Electrification is represented by the paradigm shift in electricity consumption – consumers are transformed in prosumers by selling and buying electricity and thus generating income and solar entrepreneurship.

6 CONCLUSION

Our analysis provides a historic overview of the Bangladesh solar electrification program. The public entity IDCOL has been proven to be highly effective as a single aggregator for financing, standardizing and deploying SHSs on large scale in rural Bangladesh. Based on their policies, the key challenges of last mile distribution – financing as well as delivery of affordable, but high-quality SHSs combined with effective customer service – could be solved. The program's policies combine an attractive ownership model of the SHS for the end-user, affordable consumer loans as well as a highly enforced and developed local distribution and domestic supply side, which are incentivized by subsidies and soft loans. A local solar industry has emerged. Local manufacturing and distributors employ more than 100,000 people nowadays. Our institutional analysis identifies a well-structured division of the dimensions production, distribution and financing between the multiple stakeholders. The public sector, represented by the infrastructure agency IDCOL, is coordinating the policy level and managing the financial dimension of the program, while the production is in the hands of private manufacturing company. The provision is organized as a hybrid public private partnership model, which is characterized by MFIs, NGOs and social for-profit enterprises. In regards to swarm electrification, the facilitation of grid-like infrastructure for end-users should not be perceived as a competition to the national utility and the POs, but rather as an opportunity – on the one-hand, POs can tap into their existing SHS installations and stimulate consumption, productive use and entrepreneurship by repurposing and reallocating electricity usage at community level. On the other hand, as the P2P swarm microgrids grow towards the national grid, they can become the optimal service provider; the surplus electricity of each standalone solar system could be traded by the participants in the swarm grids.

REFERENCES

[1] Khandker, S.R., Samad, H.A. Sadeque, Z.K.M., Asaduzzaman, M., Yunus, M. & Haque, A.K.E., *Surge in Solar-Powered Homes*, World Bank: Washington, 2014.

[2] Bangladesh Bureau of Statistics, *Report of the Household Income & Expenditure Survey 2010*, Bangladesh Bureau of Statistics-Ministry of Planning: Dhaka, 2011.

[3] World Energy Outlook Team, *Energy Access Outlook 2017: From Poverty to Prosperity,* International Energy Agency: Paris, 2017.

[4] REN21, *Renewables 2017 – Global Status Report,* Renewable Energy Policy Network for the 21st century: Paris, 2017.

[5] World Bank, *Tracking SDG7: The Energy Progress Report*, World Bank: Washington, 2018.

[6] Briceño-Garmendia, C., Smits, K. & Foster, V., *Financing Public Infrastructure in Sub-Saharan Africa*, World Bank: Washington, 2009.

[7] Beckers, T., Gizzi, F. & Jäkel, K., Ein Untersuchungsansatz für Systemgüter: Einordnung, Darstellung, Vorgehen bei der Anwendung. *TU Berlin/WIP-Working Paper,* **2012–01**, 2012.

[8] Hassheider, H., *Die Bereitstellung Überregionaler Straßeninfrastruktur – Eine Institutionenökonomische Analyse der Optimalen Organisationsform und -größe*, Vandenhoeck & Ruprecht: Göttingen, 2005.

[9] Ewers, H.-J., *Privatisierung der Bundesautobahnen*, Vandenhoeck & Ruprecht: Göttingen, 1995.

[10] Ménard, C. & Saleth, R.M., The effectiveness of alternative water governance arrangements. *Towards a Green Economy,* ed. M. Young, Routledge/United Nations Environment Program: Paris, 2012.

[11] Ménard, C. & Shirley, M.M., *Handbook of New Institutional Economics*, Springer: Dordrecht, 2005.

[12] Hankel, L., Finanzierungs und betreibermodelle für netzferne elektrizitätsversorgungssysteme mit photovoltaik: Institutionenökonomische analyse in subsahara-Afrika. *TU Berlin/WIP-Working Paper*, 2014.

[13] Klatt, J.P., *Eine Institutionenoekonomische Analyse von Finanzierungslösungen für die Bundesfernstraßen*, Nomos: Baden-Baden, 2011.

[14] Wealer, B., Seidel, J.P. & Hirschhausen von, C., Decommissioning of nuclear power plants and storage of nuclear waste. *The Technological and Economic Future of Nuclear Power,* eds R. Haas, L. Mez & A. Ajanovic, Springer VS: Wiesbaden, 2019.

[15] Martinot, E., Cabraal, A. & Mathur, S., World Bank/GEF solar home system projects: experiences and lessons learned 1993–2000. *Renewable and Sustainable Energy Reviews*, **5**(1), pp. 39–57, 2001.

[16] World Bank, *Project Appraisal Document. Rural Electrification and Renewable Energy Development,* World Bank: Washington, 2002.

[17] Sadeque, Z., Rysankova, D., Elahi, R. & Soni, R., *Scaling Up Access to Electricity: The Case of Bangladesh,* World Bank: Washington, 2014

[18] Asaduzzaman, M., Yunus, M., Haque, A.K.E, Azad, A.A.M., Neelormi, S. & Hossain, M.D.A., *Power from the Sun: An Evaluation of Institutional Effectiveness and Impact of Solar Home Systems in Bangladesh,* Bangladesh Institute of Development Studies/World Bank: Dhaka, 2013.

[19] Bertsch, N. & Marro, P., Making RE a success in Bangladesh: Getting the business model right. *ADB South Asia Working Paper Series*, **41**, pp. 1–39, 2015.

[20] Tiedemann, S., The Bangladeshi solar home system (SHS) Programme: Relevance and development of the supply side. *CISD Yearbook of Global Studies*, **2**, pp. 214–243, 2015.

[21] IRENA, *Renewable Energy and Jobs: Annual Review 2016*, International Renewable Energy Agency: Abu Dhabi, 2016.

[22] IRENA, *Renewable Energy and Jobs: Annual Review 2017*, International Renewable Energy Agency: Abu Dhabi, 2017.

[23] IRENA, *Renewable Energy and Jobs: Annual Review 2018*, International Renewable Energy Agency: Abu Dhabi, 2018.

[24] Sanyal, S., Prins, J., Visco, F. & Pinchot, A., *Stimulating Pay-As-You-Go Energy Access in Kenya and Tanzania: The Role of Development Finance,* World Resources Institute: Washington, 2016.

[25] Wimmer, N., *Green Energy for a Billion Poor: How Grameen Shakti Created a Winning Model for Social Business,* MCRE-Verl.: Vaterstetten, 2012.

[26] Tyabji, N. & Chase, J., *IDCOL-lapse: the unmaking of an Off-Grid Solar Market,* Bloomberg New Energy Finance: New York, 2016.

[27] Kebir, N., Spiegel, N., Schrecker, T., Groh, S., Scott, C. & Aliaga Ferrufino, G., *Exploring Energy SME Financing in Emerging and Developing Countries,* Sustainable Business Institute: Oestrich-Winkel, 2013.

[28] UN, UN awards US $1 million for Bangladesh solar entrepreneurship initiative to aid rural electrification, United Nations Department of Economic and Social Affairs. www.un.org/development/desa/en/news/sustainable/2017-energygrant-winners.html. Accessed on: 19 May 2019.

COMPARISON OF POTENTIAL MATERIALS FOR PRODUCER GAS WET SCRUBBING IN PILOT-SCALE GASIFICATION UNIT

JAKUB ČESPIVA, JAN SKŘÍNSKÝ & JÁN VEREŠ
VŠB – Technical University of Ostrava, Czech Republic

ABSTRACT

The wet scrubber technology for tar content decrease can bring a necessary cleaning solution for the biomass gasification product syngas. This paper investigates and compares several materials suitable for a tar capture. The scrubbing liquids were easily accessible and economically affordable liquids, some of which are known to possess a solving effect on the tar compounds. There are two organic chemicals: acetone and propane-2-ol, demineralised water and, as a representative of oily substances, rhodorsil oil. The investigation was divided in two parts. The first part was a producer gas partial online sampling through a set of five impinger bottles, each filled with 100 ml of the investigated material. In the second part of the investigation, all samples underwent a laboratory analysis, the results of which had given information about the trapped tar content after the sampling. The analysis consisted of weight difference, ultimate tar content and tar solvency activity examining. Of the tested scrubbing liquids, acetone showed a good capturing activity of tar, which was around 7.7 $g{\cdot}m^{-3}$. Even better results were detected in the case of propane-2-ol where 12.5 $g{\cdot}m^{-3}$ was captured. On the other hand, water showed no solving activity and the tar capture rate was significantly lower – that is, 1.2 $g{\cdot}m^{-3}$. Surprisingly, similar results occurred in the case of rhodorsil oil. The downside of the materials with good solubility is a high evaporation tendency, which may complicate their final practical usage. The presented values in this paper are related to normal conditions of wet gas (temperature of 273.15 K; pressure of 101,325 Pa).

Keywords: wet scrubbing, scrubbing liquids, tar, tar removal, gasification, biomass, comparison.

1 INTRODUCTION

The issue of global weather changes associated with anthropogenic sources of carbon dioxide and other substances is an often discussed topic worldwide. Following that, the demand for green energy and alternative fuels is ever growing. One of the important sources of alternative fuel could be biomass gasification. Its product, syngas, contains mainly CO, H_2, CH_4, C_xH_y, O_2, CO_2 (if autothermal gasification) and N_2 (if air gasification). After a separation treatment valuable synthetic gas, consisting of CO and H_2, is obtained.

However, before well-developed commercial use of gasification units, several technical issues have to be resolved. One such issue is the formation of pollutants, contained within the producer gas. Among these pollutants, tar compounds above all cause critical technical complications during the operation which could lead to an undesired shutdown of the technology.

Formation of tar compounds represents one of the greatest difficulties during the gasification process of the pilot small-scale experimental gasification facility at the Energy Research Center (ERC), VSB – Technical University of Ostrava, Czech Republic. These tar compounds are produced in 0.3–10.5 $g{\cdot}m^{-3}$ amounts during the operation of the unit [1].

Chłond [2], in his work, finds tar formation of BTX (benzene, toluene and xylene) at a rate of 10.1 $g{\cdot}m^{-3}$ and heavier tars at a rate of 10.4 $g{\cdot}m^{-3}$ during the gasification of spruce pellets. While the gasification temperature was equal to 750°C, the overall tar formation rate was over 20 $g{\cdot}m^{-3}$, according to Chłond [2]. Similar results were obtained in other samplings,

WIT Transactions on Ecology and the Environment, Vol 237, © 2019 WIT Press
www.witpress.com, ISSN 1743-3541 (on-line)
doi:10.2495/ESUS190081

where a mixture of residue derived fuel (RDF) and wood pellets was used at the same temperature.

Heavy hydrocarbons, primarily defined as tar, officially include all the condensable organic hydrocarbons which have a higher dew point (or alternatively higher molecular weight) than benzene [3]. Fig. 1 shows a typical composition of the biomass tar compounds. The values are approximate as the composition varies depending on the gasification conditions and fuel properties [3], [4].

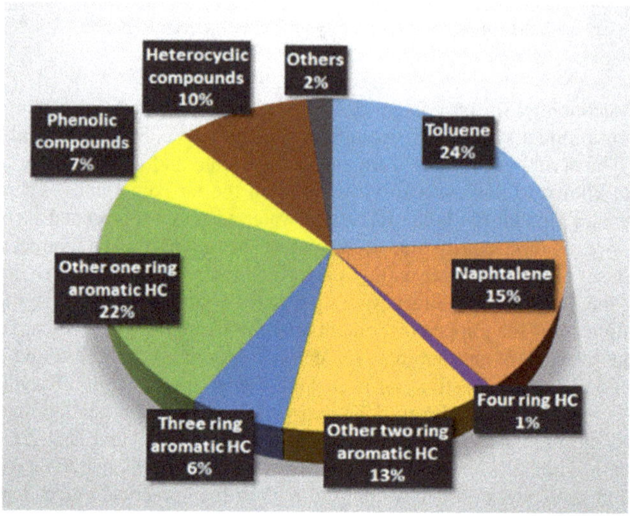

Figure 1: Typical composition of biomass tars [4].

The large amount of tar definitions and measurement methods, as well as the wide spectrum of organic compounds, makes it almost impossible to capture tars with a clear definition. There are hundreds of tar compounds that have significant differences between them. But a main categorization can be adopted according to their principal characteristics, their detectability and the number of aromatic rings in their molecules. The Energy Research Centre of The Netherlands (ECN) defines the tar classes as seen in Table 1 [5], [6].

Tars can be corrosive and can cause clogging and disrupt the operation of pyrolysis and gasification plants [3]. As the tar is a highly viscous matter, solid particles can adhere to the surface and speed up the clogging process. Another reason to clean the producer gas of all tar compounds is its further usage in the ERC Fischer-Tropsch liquefaction synthesis technology, which is very sensitive to impurities of any form. Therefore, the removal of tar is crucial also from this point of view.

There are many ways to remove these environmentally hazardous pollutants using primary and secondary tar removal methods, according to the placement. While the primary methods decrease tar production itself, the secondary methods deal with tars already produced. Ahrenfeldt and Knoef [7] categorise secondary pollutants´ removal methods as: cyclone filters, barrier filters, electrostatic precipitators (ESP) and wet scrubbers. For tar removal specifically it is wet scrubbers, ESP and catalytic conversion (primary or secondary). Both ESP and catalytic conversion tar decomposition approaches require significant technological improvements, investments and space demands to any existing or brand new gasification facility.

Table 1: Tar classification [5].

Class	Type	Description	Examples
1	Gas chromatography undetectable	–	Biomass fragments
2	Heteroclytic compounds	These are components that generally exhibit high water solubility.	Phenol, cresol, quinoline, pyridine
3	Aromatic components	Light hydrocarbons, which are important from the point view of tar reaction pathways, but not in particular towards condensation and solubility.	Toluene, xylene, styrene ethylbenzene (excluding benzene)
4	Light polyaromatic hydrocarbons (2-3 rings PAHs)	These components condense at relatively high concentrations and intermediate temperatures.	Naphthalene, indene, biphenyl, anthracene
5	Heavy polyaromatic hydrocarbons (>3 rings PAHs)	These components condense at relatively high temperature at low concentrations.	Fluoranthene, pyrene, crysene
6	GC detectable, not identified compounds	–	Unknowns

The wet scrubber solution provides a physical removal of tar compounds using a scrubber unit working with ideal scrubber liquid. In the scrubber, particles are collected by a collision with droplets of the scrubber liquid [7]. Such liquid could be demineralised water, but also other, lipophilic material with solving effect such as those which were tested in this paper in order to be used with better overall removal efficiency.

Nakamura et al. [8] claims 96% tar removal efficiency using bio-oil scrubber coupled with char filtration section. Oil alone, if new, would work with 64.5% efficiency at a temperature of 50°C. Other authors describe venturi scrubber technology to remove tar from the producer gas with 50–90% removal efficiency. Often used material for such technology in industry is a rapeseed methyl-ester (RME) oil [9], [10].

The downside of scrubbing technology (e.g. water scrubbing) is the sorbent contamination. The waste liquid cannot be discharged to the environment without chemical and/or biological waste water treatment and usually requires additional treatment facilities. An interesting solution is the warm gas clean process, which operates with temperatures above the scrubber liquid dew point but below the condensation temperature of most tars. This provides lower scrubber material cost and less intensive waste liquid treatment [7], [8], [11].

2 MATERIALS AND METHODS

2.1 Description of gasification regime and conditions

The gasification process was realized in a fixed bed counter-current reactor. The regime of the process was autothermal; the gasification reactor obtained the heat for the chemical reactions directly from the partial combustion reactions themselves instead of an outer heat source. The gasification media was ambient air without preheating.

The reactor is a small-scale pilot unit working with power up to 100 kW. The shape of the grate is circular, designed for wood pellets of a diameter around 5 mm and length no more than 30 mm. These pellets were used as a fuel in this experiment. From the fuel storage they were led and piled above the grate by two synchronized screw conveyors (horizontal and vertical). The fuel consumption was approximately 40 $kg \cdot h^{-1}$. The producer gas was generated with an average flow of 56 $m^3 \cdot h^{-1}$. During the whole experiment, the temperature within the reactor was held in a range from 700 to 800°C. This temperature range is typical for producing high amount of polluting tar particles in this reactor.

The examples of the chemical composition of producer gas for Rhodorsil oil and Acetone are shown in Figs. 2 and 3, respectively. The overall average values of the dry producer gas composition during the experiment was H_2 7%, CO 22%, CO_2 9%, CH_4 4.5%, CnHm 0.5%, O_2 2%, N_2 55%, according to the data from portable gas analyser [1], [2].

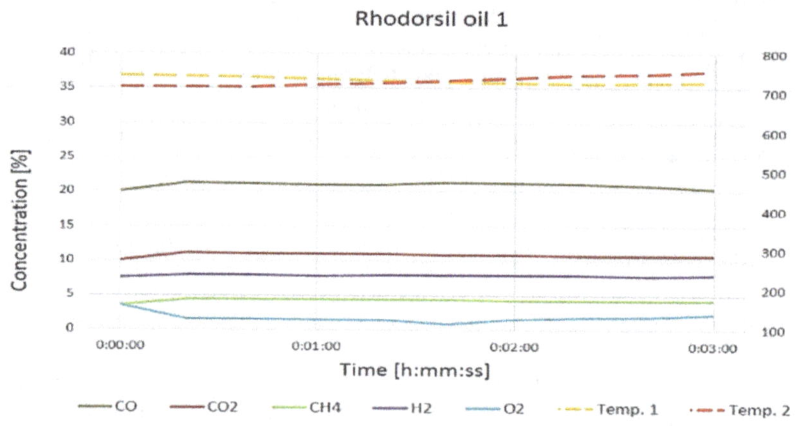

Figure 2: Producer gas chemical composition example – Rhodorsil oil 1.

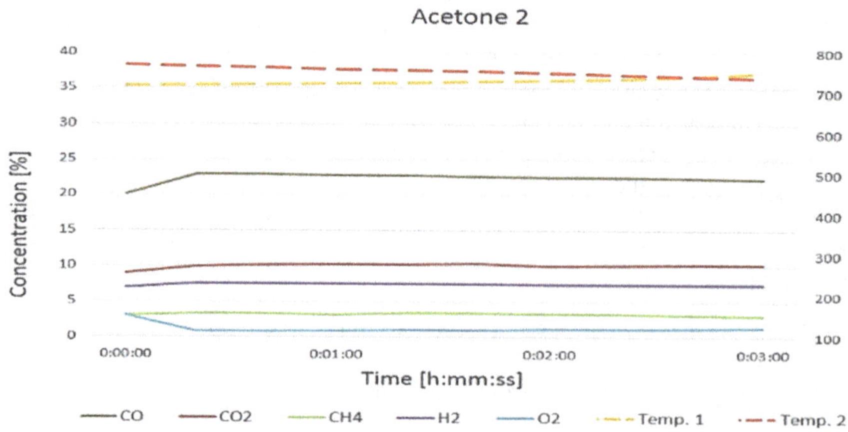

Figure 3: Producer gas chemical composition example – Acetone 2.

2.2 Scrubbing liquids description

As mentioned above, among the tested cleaning materials for sufficient tar removal, there were acetone (dimethyl ketone), propan-2-ol (also known as isopropyl alcohol, isopropanol or 2-propanol), rhodorsil oil based on siloxane chains and demineralised water. Acetone, as well as propan-2-ol, is known for good cleaning activity of tar stained surfaces. Their polarity is close to that of most organic compounds which makes them proper tar solvents.

Oils are materials often used in wet scrubber technology for flue gas treatment and so the effect of rhodorsil oil on tar treatment was investigated in this paper. The last liquid to be tested in comparison was demineralised water. The solubility of tars in water is very low, but because water is easily accessible, it is also included among the scrubbing liquids. A detailed description of the scrubbing liquids and their properties is listed in Table 2.

Table 2: Properties of scrubbing liquids for 20°C.

Material	Purity (%)	Chemical formula	Molar mass ($g \cdot mol^{-1}$)	Density ($kg \cdot m^{-3}$)	Boiling point (°C)
Acetone [13]	99.5	C_3H_6O	58.08	786	56
Propane-2-ol [14]	99.5	C_3H_8O	60.10	785	82
Demi. water [15]	100.0	H_2O	18.02	998	100
Rhodorsil oil [16]	100.0	–	–	959	280**

*25°C; **flash point.

2.3 Sampling

The sampling of the gas was realized through a sampling point situated on the pipeline right after the reactor. The suction rate was equal to 11 $l \cdot min^{-1}$. The sampling was done by a heated probe with a ceramic filter to capture solid particles from the producer gas. After the probe followed a set of five impinger bottles filled with the tested liquid. Each bottle contained 100 ml and was emerged in water/salt/ice mixture to keep the ambient temperature close to 0°C.

Each sampling was terminated after an interval of 3 minutes and bottle contents were mixed together. This was done twice for each material, making eight 0.5 l samples (two for each material) suitable for further laboratory investigation.

After the sampling impinger bottles train, another set of similar bottles followed. Those were filled with additional cleaning liquid and placed in the freezing environment in order to protect the membrane pump, flowmeter and chemical composition analyser Pollutec Gas Analyser 3000p. This analyser works with 1 $l \cdot min^{-1}$ and detects all important components from the analysed gas using a dual beam nondispersive infrared detector (NDIR), a thermal conductivity detector (TCD) and an electrochemical detector (ECD). The schematic diagram of the sampling track is shown in Fig. 4. The sampling train was set and installed according to CEN technical specification for tar sampling [12].

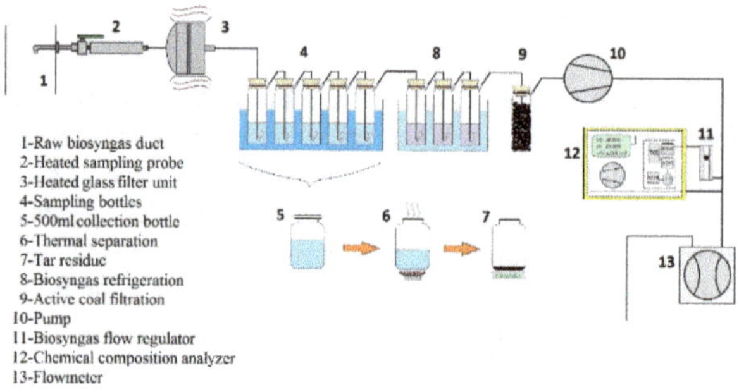

Figure 4: Tar sampling track.

2.4 Tar content investigation

Prior to the sampling, the weight of the filled bottles was checked on Kern DE 12K1N measuring device and recorded for later weight difference analysis. After each experiment period of 3 minutes, the sampling probe was closed and the pump stopped. The sampling impinger bottles were then detached from the train in order to maintain the weight difference analysis. This analysis provided information about weight change associated with increased tar content trapped within the cleaning material.

In the second step, the collection bottle with the tested liquid was attached to a laboratory distillation column where the examined material was separated to leave only the residue tar behind.

In the last step to investigate the sufficiency of the cleaning materials, their solubility was examined. The tar solubility test consisted of basic tar/solvent mixing. During the test, the exposition of visible tar matter within the scrubbing liquid was measured. The time needed for a complete solvency of tar matter is directly proportional to the solubility of the material.

3 RESULTS AND DISCUSSION

3.1 Results

The investigation of tar capture ability of chosen liquid materials consisted of three separated parts; weight analysis, evaporation tar residue analysis and tar solvency analysis. All materials' investigations were done in this order: Acetone 1 – propan-2-ol 1 – demineralised water 1 – rhodorsil oil 1 – acetone 2 – propan-2-ol 2 – demineralised water 2 – rhodorsil oil 2.

The results of weight records were unexpected. Demineralised water and rhodorsil oil have shown a weight increase of 1.3 $g \cdot l^{-1} \cdot min^{-1}$ on average. In the case of solvent materials, a weight decrease was observed. The weight of propan-2-ol was lower by 1.4 $g \cdot l^{-1} \cdot min^{-1}$ on average, while that of acetone dropped by 16 $g \cdot l^{-1} \cdot min^{-1}$ on average. This phenomenon can be explained by evaporation of the materials, which tend to evaporate at lower temperatures than water or oil (see Table 2).

This theory was proved by examining the content increase in additional refrigeration bottles, following after the sampling train. Both water and oil samples did not change the volume of the freezing impinger bottles' content. The problem with the evaporation could be solved by condensation surfaces, that is, mist eliminators, placed above the scrubber to avoid evaporated material to escape the unit.

According to materials safety data sheet (MSDS), both acetone and propan-2-ol are considered non-hazardous materials for the environment and o-zone layer with good biodegradability in water and soil. They do not cause a greenhouse effect and are recycled by distillation. However, it is in our best interest to keep away from possible environment contamination and from contact with human skin, eyes and respiratory tract where it may cause serious irritation. Another reason to apply mist eliminators is the exposure limits which are 2,500 mg·m^{-3} in the case of acetone and 980 mg·m^{-3} in the case of propan-2-ol, according to the Occupational Safety and Health Administration (OSHA). Acetone and propane-2-ol are flammable and explosive liquids and their usage is temperature limited [13], [14].

Due to insufficient results obtained from weight analysis, a more detailed investigation of tar content was necessary. This was done with a distillation column, part of which was a water bath with adjustable temperature. Each of the 500 ml samples was in the rotating test-glass, emerged in the water bath and connected to the distillation column. Vapours then condensed in this water-cooled column and leaked off. The result of this process was a separated tar matter within the test-glass, the weight of which was analysed. The distillation of the materials was held in two steps.

In the first step, water bath temperature was adjusted below the material boiling point to separate lighter tars like cyclohexane. Then, the temperature was raised above the boiling point of the material to evaporate it. After the evaporation of the scrubbing liquid, only heavier tars remained within the test-glass. According to Table 2, the bath temperature was set above 57°C to evaporate acetone, 83°C to evaporate propane-2-ol and 101°C to evaporate water. The separation of rhodorsil oil in this way was not possible in the lab-scale conditions.

The last test was aimed on the tar solubility of the scrubbing liquids. In this test, 0.05 g of very fine and homogenised fraction of the solid tar, collected from the technology pipeline, was solved in 150 ml of each material. The mixture was gently stirred and the time, needed for a complete visible dissolution of the tar, was measured. The fastest dissolution was measured in acetone – 60 s approximately. Propan-2-ol solved the tar powder five times slower while demineralised water and rhodorsil oil had no visible solving effect on the tar.

In addition, a 50% mixture of water and acetone was examined as well. Such mixture is also a potential solution for the tar removal. The result of approximately 180 s confirmed that water behaves as tar solvency restraint but could positively affect the evaporation disadvantage of acetone or propan-2-ol. Fig. 5 shows the samples examined for tar solving effect. The results of all three analyses are presented in Table 3.

Figure 5: Tar solving samples. (A) Acetone; (B) Acetone/water; and (C) Water.

Table 3: Average results of the comparing analyses.

Material	Weight analysis	Tar analysis	Solubility analysis (0.05 g in 150 ml)
	(g)	($g \cdot m^{-3}$)	(s)
Acetone	-16.0	7.7	60
Propane-2-ol	-1.4	12.5	300
Demin. Water	1.3	1.2	>1,600
Rhodorsil oil	1.3	<1.3	>1,600

3.2 Discussion

At this point it is necessary to mention several important factors which may, in some way, affect the results of these experiments. First of all, gasification is a complex process, much dependent on many circumstances including temperature, pressure, oxidant activity, fuel properties etc. Especially in the case of fixed bed reactors it is difficult to keep the process stable and in desired deviation throughout long periods of time. The producer gas content may change and the scrubber technology demands may vary during the operation.

Other issues concern the sampling and scrubbing liquids; the efficiency of scrubbing materials decreases in time. During the scrubbing process, particles and solid contaminates would be accumulated in the absorbent, where these impurities obstruct the absorption mechanism, leading to decreased tar removal efficiency [10]. Therefore, a regeneration of the material would be necessary and must be taken into consideration in order to choose the most suitable scrubbing material.

According to previous studies, oils are often used materials for wet scrubbing to clean the producer gas. In the case of rhodorsil oil, the results were not better than in the case of other tested liquids, however, the analysis was inaccurate due to the physical properties of this oil.

Apart from technical oils, some authors successfully tested bio, vegetable (e.g. soybean or canola oil) and waste-cooking oils, however, polluted oil treatment is still a big issue for the gasification technology. Evaporable material with boiling point below that of tars could bring an interesting solution to the tar removal problems. Such approach would turn a downside of acetone or propan-2-ol into an advantage and have smaller impact on the environment, but the economical factor would remain an issue [8], [17].

As seen in Table 3, acetone and propane-2-ol are reliable solvents and absorbers of the tar compounds but with a tendency to evaporate while working in hot environment. This problem must be sufficiently solved technically in order to use these materials for producer gas scrubbing. A vapour condensing unit and evaporation itself must be examined in future studies, as well as usage of other materials or their combinations. This could lead to better overall and economic efficiency of the tar removal as acetone and propan-2-ol are both promising materials but could hardly work in pure state.

4 CONCLUSION

The tar removal from gasification generated producer gas, using four different scrubber liquids, was investigated and compared in this paper. The investigation was held through three-step analysis. The results have shown the complexity of acetone and propan-2-ol usage caused by their high evaporation rate. On the other hand, these materials proved to be very good solvents of tar compounds and to be capable of absorbing them well, up to ten times better than rhodorsil oil or demineralised water. However, for practical usage of acetone or propan-2-ol in wet scrubber technology, a demister or condensing unit must be applied in order to avoid exceeding evaporation.

The subject of the following study will focus on a combination of scrubbing liquids for obtaining the best results. The results of the analyses will serve as a basis for further technological proposal for wet scrubber technology construction in a pilot-scale biomass gasification unit in ERC.

ACKNOWLEDGEMENTS

This work was prepared within the project "Innovation for Efficiency and Environment – Growth", identification code LO1403 with the financial support from the Ministry of Education, Youth and Sports in the framework of the National Sustainability Programme I. This article was also prepared within the project SP2019/89 "Research of conversion processes and utilization of waste heat in the framework of fuel-energy technological complexes".

REFERENCES

[1] Ochodek, T. & Najser, J., *Zplyňování biomasy I*, Ostrava: Výzkumné energetické centrum, pp. 74–80, 2014. (In Czech.)

[2] Chłond, R.A., *Zvyšování výhřevnosti a kvality plynu ze zplyňování biomasy*, PhD thesis, Ostrava: VŠB – Technical University of Ostrava, 2012. (In Czech.)

[3] Neeft, J.P.A., Knoef, H.A.M. & Onaji, P., *Behavior of Tars in Biomass Gasification Systems: Tar Related Problems and Their Solutions*, Novem EWAB Program Report 9919, MHP Management Services, 1999.

[4] Shen, Y. & Yoshikawa, K., Recent progresses in catalytic tar elimination during biomass gasification or pyrolysis – A review. *Renewable and Sustainable Energy Reviews,* **21**, pp. 371–392, 2013.

[5] Bergman, P.C.A., Paasen, S.V.B. & Boeriggter, H., The novel "OLGA" technology for complete tar removal from biomass producer gas. *Pyrolysis and Gasification of Biomass and Waste*, 2002

[6] Vakalis, S., Patuzzi, F., Moustakas, K., Sotiropoulos, A., Malamis, D. & Baratieri, M., Analysis of tar compounds and quantification of naphthalene from thermal treatment of household biowaste. *Journal of Environmental Management,* **216**, pp. 153–159, 2018.

[7] Ahrenfeldt, J. & Knoef, H.A.M., *Handbook Biomass Gasification*, BTG Biomass Technology Group: Enschede, 2005.

[8] Nakamura, S., Kitano, S. & Yoshikawa, K., Biomass gasification process with the tar removal technologies utilizing bio-oil scrubber and char bed. *Applied Energy*, **170**, pp. 186–192, 2016.

[9] Hasler, P. & Nussbaumer, T., Gas cleaning for IC engine applications from fixed bed biomass gasification. *Biomass and Bioenergy,* **16**(6), pp. 385–395, 1999.

[10] Unyaphan, S., Tarnpradab, T., Takahashi, F. & Yoshikawa, K., An Investigation of Low Cost and Effective Tar Removal Techniques by Venturi Scrubber Producing Syngas Microbubbles and Absorbent Regeneration for Biomass Gasification. *Energy Procedia,* **105**, pp. 406–412, 2017.

[11] Woolcock, P.J. & Brown, R.C., A review of cleaning technologies for biomass-derived syngas. *Biomass and bioenergy,* **52**, pp. 54–84, May 2013.

[12] CEN Technical Specification, Biomass Gasification – Tar and Particles in Producer Gases – Sampling and Analysis, prepared by CEN task force BT/TF 143 WI CSC 03002.4TC, 2005.

[13] LabChem. Materials safety data sheet: Acetone. www.labchem.com/tools/msds/msds/LC10420.pdf. Accessed on: 1 Mar. 2019.
[14] LabChem. Materials safety data sheet: Isopropyl Alcohol (2-Propanol). www.labchem.com/tools/msds/msds/LC15750.pdf. Accessed on: 1 Mar. 2019.
[15] LabChem, Materials safety data sheet: Isopropyl Alcohol (2-Propanol). www.labchem.com/tools/msds/msds/LC26750.pdf. Accessed on: 1 Mar. 2019.
[16] Rhodorsil Oils 47 V 50 až 47 V 1000; Technical list; February 1996. (In Czech.) www.stavochemie.cz/tl/tl-rhodorsils.pdf.
[17] Paethanom, A., Nakahara, S., Kobayashi, M., Prawisudha, P. & Yoshikawa, K., Performance of tar removal by absorption and adsorption for biomass gasification. *Fuel Processing Technology,* 104, pp. 144–154, 2012.

PRODUCTION OF ANTIOXIDANTS FOR BIODIESEL FROM STRAW BLACK LIQUOR DEPOLYMERIZATION

JOSÉ PALOMO[1], SAMUEL MOLES[1], JESÚS SALAFRANCA[2], NOEMI GIL-LALAGUNA[1], ALBERTO GONZALO[1] & JOSÉ LUIS SÁNCHEZ[1]
[1]Thermochemical Processes Research Group, Universidad de Zaragoza, Spain
[2]University Analytical Research Group, Aragon Institute of Engineering Research, Universidad de Zaragoza, Spain

ABSTRACT

Black liquor produced from semi-chemical pulping of barley straw has been depolymerized at different temperatures (250 to 325°C), times (1 to 3 h) and initial solids content in the black liquor (7 to 14%). The obtained degraded and soluble lignin was then used as biodiesel stabilizer (against oxidation). The depolymerization products have been analyzed by gel permeation chromatography, gas chromatography-mass spectrometry (GC-MS), Fourier-transform infrared spectroscopy (FTIR), and phenolic content was quantified by using the Folin–Ciocalteu method. Regarding the molecular weight distribution, temperature and time have been the most influential factors, especially temperature, which decreased the observed molecular weight. The presence of volatile phenolic compounds has been determined by GC-MS. Temperature was found to be the most influential factor: the higher the temperature, the lower the concentration of volatile phenolics. Regarding the FTIR, the measured spectra hardly change when modifying the experimental conditions. Oxidation stability of biodiesel increased with the reaction temperature applied to black liquor, suggesting that the increase of oxidation stability is not only caused by the volatile phenolic compounds.

Keywords: black liquor, lignin, biodiesel, oxidation, additives.

1 INTRODUCTION

Straw is a readily available source of cellulosic material for the pulp and paper industry, especially for the boardmaking industry [1]. Black liquor is a byproduct, which contains lignin, hemicellulose and other compounds including the cooking chemicals. The recovery of straw black liquor cannot be accomplished by the same burning process as wood kraft black liquors, mainly due to its silica content which causes high viscosity at low solid concentration and scaling problems in the evaporators and burning chambers [2]. Low density and seasonal production are also other disadvantages that have reduced the use of these materials during the last years in favor of wood pulping [1].

Nevertheless, using fiber sources other than wood is interesting from an environmental point of view, as the use of these materials could lead to a better forest management, and provide a more profitable use of agricultural residues. Specifically, soda semi-chemical pulping is a process that uses no sulfur in the digestion, and can render a pulp with good mechanical properties that can be mixed with secondary (recovered) fiber for cardboard paper production [3], [4].

As an important by-product in any pulping process, black liquor contains a significant fraction of the plant lignin, which is the most abundant source of renewable phenolic compounds in nature. Therefore, depolymerization of the lignin macromolecule can be an alternative to combustion, allowing a valorization of its unique chemical structure. Lignin depolymerization leads to the formation of several products including gases, monomers (monoaromatic compounds), oligomers (polyaromatic compounds), non-converted lignin and condensation products. However, because of its stable and polymeric nature, it is difficult to obtain a single product by depolymerization.

WIT Transactions on Ecology and the Environment, Vol 237, © 2019 WIT Press
www.witpress.com, ISSN 1743-3541 (on-line)
doi:10.2495/ESUS190091

In this work, depolymerization of the lignin contained in a stream of straw black liquor has been studied aiming at converting it to antioxidant additives for biodiesel. Oxidation stability is among the monitored parameters for its commercialization, being poor oxidative stability one of the major issues to deal with before using biodiesel as a fuel for diesel engines. Hence, the main goal of this work is to evaluate the antioxidant potential of lignin-derived products obtained from black liquor depolymerization, as well as to characterize these products using different analytical techniques.

2 MATERIALS AND METHODS

2.1 Black liquor production

Black liquor (BL) was obtained from batch cooking barley straw for 3 hours, using a liquid to straw weight ratio of 10:1 and a NaOH weight of 9.5% over the straw dried weight. Temperature was kept at 98–100°C during the whole process. Details of the procedure can be found in the published report by Marín et al. [3]. After the digestion, the solid pulp was filtered and discarded, keeping the resulting black liquor for the depolymerization experiments. After the digestion, BL was further concentrated by evaporating water up to a 15 wt.% solid content, which includes lignin and ash (NaOH among other compounds). This BL was kept in a freezer and, prior to each experiment, was diluted to the desired concentration in distilled water. Table 1 shows the black liquor solid (BLS) characterization.

Table 1: Black liquor solid characterization.

Proximate analysis (wt.%, as received)		
Moisture:	2.1±0.5	EN ISO 18134:2015
Ash:	34.8±0.4	EN ISO 18122:2015
Volatiles:	55.2±0.3	EN ISO 18123:2015
Fixed carbon:	7.9	By difference
Ultimate analysis (wt.%, as received)		
C:	36.5±0.4	Elemental Analyzer LECO CHN 628 Series
H:	5.07±0.05	
N:	0.82±0.04	
O:	22.8	(by difference)
HHV (MJ/kg)	14.2±0.1	C2000 IKA bomb calorimeter

2.2 Black liquor depolymerization and biodiesel oxidation stability testing

Depolymerization experiments were carried out in a PARR Autoclave reactor, model 4575A. BL was treated varying the initial solids content in BL from 7 to 14%, reaction time from 1 to 3 h and temperature from 250 to 325°C. NaOH is known to be a catalyst in the lignin depolymerization [5], [6], thus no other catalyst was used as it was already present in the BL. In each experiment, a total mass of 200 g (water + BLS) was used. Table 2 summarizes the experimental conditions, which correspond to a 2^k factorial experimental design, with three replicates of the central point. By analysis of variance (ANOVA), the effect of the three experimental factors (time, temperature and initial BLS content) on several response variables has been analyzed.

After the reaction time, the reactor was cooled by means of a water coil. Once the system reached room temperature, residual pressure inside the reactor was released by collecting the gas product in a Tedlar bag for GC-TCD analysis (Agilent 3000 Micro Gas Chromatograph). The gas yield was then calculated using the following data: final pressure in the sealed reactor at room temperature, free reactor volume (300 mL) and gas volumetric composition. The liquid content in the reactor was poured into a beaker and weighted. The liquid product and the washing solution were filtrated and the insoluble solid residue (ISR) was dried at 100°C overnight and weighted. An aliquot of filtrated liquid was dried overnight (40°C) in order to determine, by weight difference, the amount of soluble degraded lignin (SDL) present in the liquid product. Finally, the product distribution was calculated as the mass yield of each product (gas, ISR and SDL) with respect to initial solid content in the raw black liquor. The soluble degraded lignin was the product to be tested as additive for biodiesel.

Sunflower biodiesel was produced at laboratory scale as previously reported [7]. Neat biodiesel and the liquid products generated from lignin depolymerization were blended at such a proportion that the initial addition of lignin-derived compounds into biodiesel was 3 wt.%. This blend was distilled under vacuum. Once the water had been removed, the mixture of biodiesel and lignin-derived compounds was magnetically stirred (30 min) and centrifuged (15 min, 4500 rpm). The insoluble fraction of additive settled to the bottom of the centrifugation tube, while the homogeneous upper phase, composed of biodiesel and soluble compounds, was carefully separated, after which it was considered as the doped sample of biodiesel. Biodiesel oxidation stability was measured with a PetroOXY equipment (Petrotest Instruments GmbH & Co. KG) according to the test method described in the EN 16091 standard. The improvement of PetroOXY time (OXY) with respect to the oxidation stability of neat biodiesel involved in each doping process was calculated.

2.3 Soluble degraded lignin (SDL) characterization

Gel permeation chromatography (GPC) was used to determine the molecular weight of the SDL. It is a well stablished technique for the determination of the molar mass of polymers (as Mw, Mn or Mp), such as lignin, although it has several drawbacks [8], as the need of derivatize the lignin in order to increase their solubility in organic solvents, which is known to affect the measured Mw [9]. The use of aqueous GPC is also possible, avoiding the derivatization step [10], [11]. Thus, aqueous phase GPC was chosen after testing the solubility of both lignin and its depolymerization products. A Waters 1515 isocratic pump with a 2414 Refractive Index detector was used, with a Waters Ultrahydrogel pre-column (6 x 40 mm) and two Waters Ultrahydrogel (7.8 x 300 mm) columns heated at 40°C. A mixture of 0.004% NaOH, 0.1% LiBr and 0.05% NaN_3 (all % by weight) was used as mobile phase at 0.8 mL/min. The liquid obtained from the reactor was filtered with a syringe filter (0.45 μm) and diluted to 1 g/L in the mobile phase prior to injection (100 μL). Besides the calculation of Mn/Mw, some authors have used the collected data to gain insight on the molecular weight distribution by deconvoluting the GPC chromatograms in peaks that can be ascribed to fractions with different Mw [12], [13]. This approach can be useful to assess the extent of BL depolymerization.

FTIR spectroscopy was also used in order to obtain information of the main SDL functional groups. Samples were dried and the solid measured in a Bruker Alfa FTIR spectrometer by attenuated total reflectance (ATR).

Volatile content of the SLD samples was analyzed by GC/MS using an Agilent 7890A GC with a 5975C MS detector. Aqueous samples were filtered (0.45 μm) and directly injected (1 μL, 250°C), using a HP-FFAP column (50 m long, 200 μm diameter, 0.3 μm film thickness) and a 1/50 split ratio. Initial temperature was 50°C for 2 min, then raised at 10°C/min until reaching 235°C, after what it was held for 21.5 min. Helium (99.999% purity) was used as carrier gas at constant flow (1 mL/min). Chromatograms were acquired in scan mode (m/z 45-240 Da) and the MS NIST 2014 library was used for peak identification. Guaiacol, one of the most abundant compounds identified, was used as reference for calibration, referring the concentrations to its response. Detection limit was 3.3 μg/g.

The Folin–Ciocalteu colorimetric method was also used to characterize the phenolic content of SDL, using gallic acid as calibration standard.

3 RESULTS

3.1 Product distribution

Table 2 summarizes the results of product distribution calculated as the mass yields of insoluble solid residue (ISR), soluble depolymerized lignin (SDL) and gas product with respect to the initial mass of BLS fed to the reactor (14–28 g of BLS).

Table 2: Product distribution after lignin depolymerization.

Exp. code #	T (°C)	Time (h)	BLS content (wt.%)	ISR yield (wt.%)	SDL yield (wt.%)	Gas yield (wt.%)
A1	250	1	7	7.0	87.0	2.9
A2	325	1	7	7.1	64.4	4.9
A3	250	3	7	1.3	77.6	3.6
A4	325	3	7	3.56	61.0	6.3
A5	250	1	14	4.3	73.5	2.5
A6	325	1	14	7.8	62.8	5.5
A7	250	3	14	11.6	74.7	3.4
A8	325	3	14	7.0	49.0	7.5
A9	287.5	2	10.5	5.0	74.0	3.2
A10	287.5	2	10.5	5.6	68.7	3.9
A11	287.5	2	10.5	5.7	69.2	4.8

The ANOVA analysis shows that the ISR yield is significantly influenced by the initial solid concentration: higher solid content leads to higher ISR yield. Although the other factors (temperature and reaction time) have not been shown as significant terms themselves, the relative impact of the initial BLS content depends on the value of these factors (the interactions between all the factors appear as significant terms). Thus, working with the highest initial solid content at lower temperature and for a longer reaction time lead to the greatest formation of insoluble solid (Fig. 1(a)). On the other hand, ISR formation is minimized by operating at the lowest initial BLS content, the lowest temperature and longer reaction time (Fig. 1(b)). All these interactions highlight the complexity of re-polymerization reactions occurring during the process, which should be minimized in order to reduce the loss of carbon in the form of such residual material.

According to the ANOVA analysis, the SDL yield has been found to be negatively affected by the increase of the three operating factors, where temperature was shown to have the strongest effect. Thus, the best conditions for increasing the additive production rate have been found to be the lowest temperature (250°C), the lowest BLS content (7%) and the shortest reaction time (1 h), which provided a SLD yield as high as 87%. The interaction between the three factors has been found to be a significant term, which means that the impact of each factor depends on the value of the others.

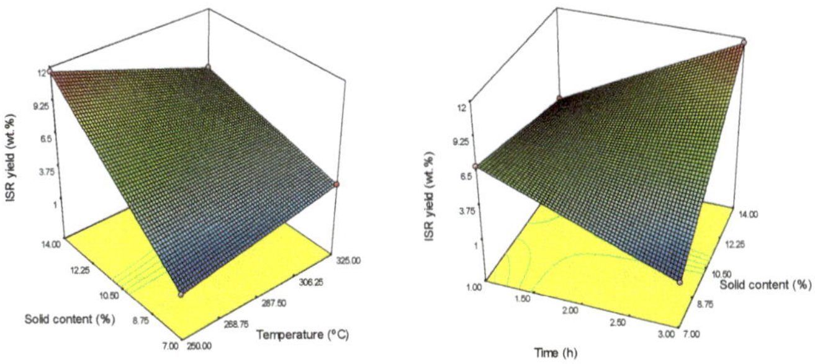

Figure 1: Effect of operating factors on ISR yield. (a) Reaction time of 3 h; (b) Temperature of 250°C.

Lastly, the gas yield accounts for 2.5–7.5% of the initial mass of solid raw material (BLS) and, according to the statistical analysis, is positively affected by temperature and reaction time. These trends agree with the aforementioned discussion: SLD production may be reduced at higher temperature and longer reaction time in favor of gas production. This gas was mainly composed of CO_2 (57–90 vol.%, N_2-free basis), CO (2–9 vol.%, N_2-free basis) and H_2 (5–39 vol.%, N_2-free basis). H_2 production was clearly enhanced at higher temperature, opposite to CO_2 fraction, which was increased at lower temperature.

3.2 GPC deconvolution

The obtained curves for both untreated (BLS) and treated black liquor (SDL) visually seem to be composed of several peaks (as shown later) that could be attributed to fractions of different molecular weights. It has been checked that all the GPC peaks can be deconvoluted in several Gaussian curves, with a good fit of the experimental GPC curve (minimum R^2 of 0.987). Assuming the same response factor (i.e., different fractions have similar refraction indexes), an analysis on how the depolymerization conditions affect the molecular weight distribution can be carried out.

The GPC profiles of the raw BLS and of a depolymerized sample (A10) are shown in Fig. 2. All the treated samples showed similar peaks. Specifically, seven peaks, which were labelled as P1 to P7, were found in all the treated samples at a similar time. Table 3 summarizes the Mp of the observed peaks in both BLS and a treated sample (central point, A10), together with their area percentage. In the case of A10, the average Mp (calculated at the average time of the 11 samples from the experimental design) and its 95% confidence interval are presented. Mathematical deconvolution of the curves allowed calculating the area percentage of each of the seven peaks. In the raw BLS, the first peak that was observed (contributing with only 3.9% of the total area) has a Mp of about 97100 Da, and completely disappears in the treated samples. The second peak, which was quite wide and centered at 20900 Da, is the most important one in area. The latter decreased after the hydrothermal treatment both in area (from 40.8 to 8.2%) and in Mp, appearing with the same shape and with a Mp of 14200 Da in the treated samples. A small peak in the raw BLS appearing at 8100 Da seemed to shift to a slightly higher Mp with a similar importance. The BLS fraction of 2400 Da increased from 6.3 to 15.4%, and also its Mp to 2520 Da. The last peak observed in BLS, which was the second by area (31.5%), increased both in area (to 58.4%) and in Mp (from 450 to 530).

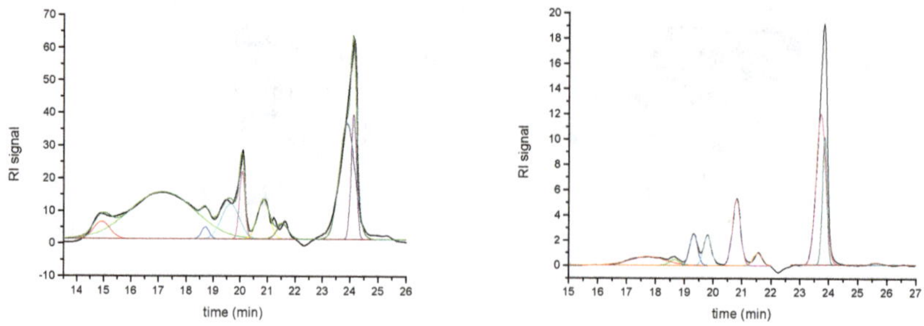

(a) (b)

Figure 2: GPC of (a) Raw BLS; and (b) Soluble degraded lignin (A10 sample).

Table 3: Mp and area percentage of deconvoluted peaks.

Raw BLS		A10 depolymerized sample	
Mp	Area percentage	Average Mp	Area percentage
97100	3.9	-	
20900	40.8	14200 ± 500 (P1)	8.2
8100	1.3	8320 ± 50 (P2)	1.6
5080	8.7	5680 ± 30 (P3)	6.8
3640	5.1	4390 ± 40 (P4)	5.8
2400	6.3	2520 ± 10 (P5)	15.4
1990	Not integrated	-	
1620	2.4	1710 ± 10 (P6)	3.3
450	31.5	530 ± 10 (P7)	58.4

Table 4 shows the average molecular weights in number (Mn) and in weight (Mw), as well as the polydispersity index (PI, calculated as the ratio of Mw to Mn) obtained by the GPC analysis.

Table 4: Average molecular weights and polydispersity determined by GPC.

T (°C)	Time (h)	BLS content (wt.%)	Mn	Mw	PI
250	1	7	3160	10260	3.2
325	1	7	2590	9350	3.6
250	3	7	2960	9230	3.1
325	3	7	2060	6960	3.4
250	1	14	3820	12390	3.2
325	1	14	2130	6620	3.1
250	3	14	3020	10290	3.4
325	3	14	1970	4990	2.5
287.5	2	10.5	3060	11040	3.6
287.5	2	10.5	2880	11020	3.8
287.5	2	10.5	2900	10330	3.6

The ANOVA of the Mn data shows that this parameter decreased when increasing both temperature and reaction time, but it was not affected by the initial concentration of solids. The effect of temperature was roughly double that the effect of time in the tested ranges. A similar trend was observed for Mw, although for this parameter the interaction between time and temperature was statistically significant. Polydispersity was not significantly affected by the experimental conditions, showing a variation from 2.5 to 3.8.

The analysis of variance (ANOVA) of the areas of the different peaks showed that the highest Mp peak (P1) decreased when temperature and time were increased. On the other hand, the smallest Mp peak (P7) increased, thus showing the effect of the depolymerization conditions.

3.3 GC/MS analysis and Folin–Ciocalteu phenolic content

Using GC/MS/FID, 30 different compounds were clearly identified in the hydrotreated liquid samples, of which eleven were phenolic-based compounds:

- The total concentration of the 11 identified phenolic compounds ranged from 530 to 2190 µg per g of treated BL. Only two of them appeared in all the samples: guaiacol (ranging from 80 to 1050 µg per g of treated BL) and phenol (from 350 to 450 µg per g of treated BL), and their joint concentration represented from 65 to 97% of the total content of phenolics. 2-methylbenzenemethanol appeared in 9 samples, but its concentration only ranged from 4.5 to 53.5 µg per g of treated BL; 2,6 dimethoxyphenol appeared in 7 samples, and in some of these samples it reached concentrations up to 600 µg per g of treated BL. These four compounds represent 75 to 98% of the total phenolics concentration in the different samples.
- Ten cyclic compounds were also identified, with only one of them appearing in all the hydrotreated samples (2,3-dimethyl-2-cyclopenten-1-one). Cycloheptanol appeared in 8 depolymerized samples. The maximum concentration for any of these compounds was 40 µg per g of treated BL.
- Four different acids were identified (acetic, butanoic, 4-hidroxybutanoic and pentanoic). Only acetic acid was present in the 11 depolymerized samples, with concentrations ranging from 4.4 to 20.6 µg per g of treated BL. Pentanoic acid, the next more abundant, was present in only 3 samples.
- Four aromatic compounds, with a maximum concentration of 50 µg per g of treated BL, where 1,2,3-trimethoxy-5-methylbenzene (appearing in 4 samples) was the most abundant.

Table 5 shows the concentration of each chemical family identified by GC/MS and the total yield of volatile phenolics, calculated from the GC/MS concentration data and expressed as mg/g BLS. The Folin–Ciocalteu (F–C) method was also used to characterize the phenolic content in SDL, and was measured as mg of gallic acid equivalent per g of SDL solid. The ANOVA calculation shows that both the initial BLS content and the temperature ("S" and "T" in equations) affected the concentration of total phenolics detected by GC/MS. Eqn 1 shows the relative influence of the factors, where temperature and BLS content are expressed as coded terms, i.e., they range from -1 for the lowest level (250°C and 7% of BLS, respectively) to +1 (325°C and 14% of BLS, respectively)

$$\text{Total phenolics concentration } (\mu g/g \text{ of treated BL}) = 1301 - 640 \cdot T + 106 \cdot S. \quad (1)$$

Table 5: Concentration of volatile compounds grouped by chemical families (μg/g treated BL), phenolic content (estimated from the F–C method) and yield of volatile phenolics.

Temp (°C)	Time (h)	BLS content (wt.%)	Acids (μg/g treated BL)	Aromatics (μg/g treated BL)	Cyclics (μg/g treated BL)	Phenolics (μg/g treated BL)	Volatile phenolics yield (mg/g BLS)	F–C (mg/g SDL)
250	1	7	8.2	107,4	43.3	1882.9	25.7	4.3
325	1	7	7.4	0,0	31.1	531.1	7.2	2.0
250	3	7	5.0	28,1	12.8	1695.7	23.0	3.6
325	3	7	24.7	n.d.	106.7	671.7	9.2	1.7
250	1	14	28.2	53.8	77.6	2189.6	14.4	5.5
325	1	14	15.4	49.1	69.9	840.5	5.4	3.2
250	3	14	38.4	92,7	85.3	1995.4	13.0	5.5
325	3	14	25.0	n.d.	68.6	602.3	3.7	1.3
287,5	2	10.5	15.5	n.d.	13.5	769.7	6.9	2.7
287,5	2	10.5	5.8	6.0	33.1	884.7	7.9	4.6
287,5	2	10.5	5.3	7.1	29.7	915.7	8.3	3.8

Temperature exerted a negative influence (the higher the reaction temperature, the lower the phenolics concentration) whereas the BLS content, as it could be expected, exerted a positive effect, that is, the higher the lignin content, more concentration of phenolics were found in the resulting liquid after hydrotreating (Fig. 3). Temperature was the most influencing factor, since its increase from 250 to 325°C led to an average decrease of phenolics of 1200 μg per g of treated BL. On the other hand, a change in BLS content from 7% to 14% only led to an increase in the phenolic concentration of approximately 200 μg per g of treated BL. Hence, the effect of the initial BLS content was not as strong as the impact of temperature, as doubling the feedstock or source of phenolics (BLS content) did not double the concentration of these compounds derived from lignin.

Fig. 3 also shows that the phenolics concentration at the central point conditions was similar to the one obtained at higher temperature (significant curvature). The ANOVA can be also applied to the measured concentration of the three most abundant compounds (guaiacol, phenol and 2,6 dimetoxy phenol). The ANOVA of the guaiacol concentration shows that only the reaction temperature exerted a significant influence (Fig. 4(a)). As expected, higher temperatures led to a lower production of guaiacol, as shown in eqn (2):

$$\text{Guaiacol concentration } (\mu\text{g/g treated BL}) = 551\text{-}407\cdot T. \quad (2)$$

On the other hand, the phenol concentration was not statistically affected by any of the experimental conditions tested, whereas 2,6-dimethoxyphenol concentration decreased with both temperature ("T") and reaction time ("t") and increased with BLS content ("S") (eqn (3); Fig. 4(b)).

$$\text{2,6-dimethoxyphenol concentration } (\mu\text{g/g treated BL}) = 252\text{-}249\cdot T\text{-}41\cdot t + 16\cdot S + 35\cdot T\cdot t. \quad (3)$$

The production of volatile phenolic compounds was also calculated in terms of mass yield with respect to the mass of black liquor solid (mg of phenolics per g of BLS) (Table 5). According to the ANOVA, both the temperature and the initial BLS content exerted a negative influence on the conversion rate of BLS to volatile phenolics, as shown in eqn (4):

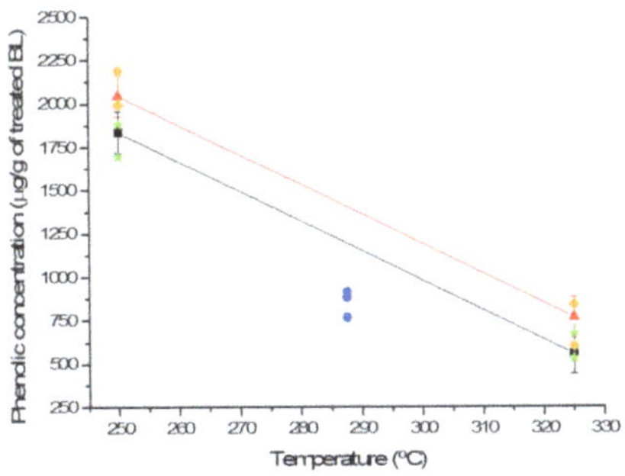

Figure 3: Concentration of phenolic compounds in the treated black liquor (experimental data: ★ 7% solids, ◆ 14% solids, ● central points).

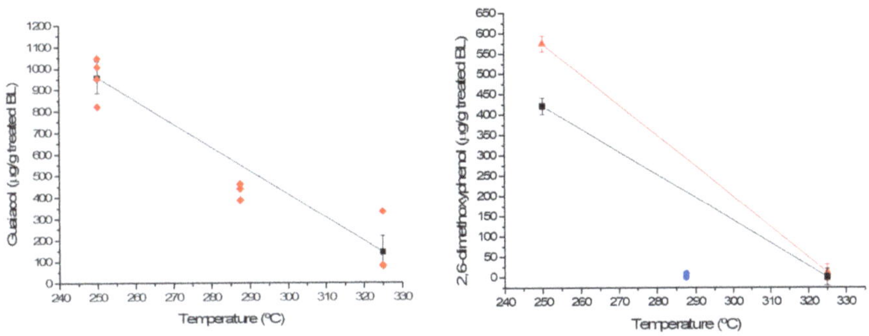

(a) (b)

Figure 4: (a) Concentration of guaiacol (◆ experimental points); (b) 2,6-dimethoxyphenol in the treated black liquor (▲ 1 h reaction time, ■ 3 h reaction time, ● central points).

$$\text{Volatile phenolics yield (mg/g BLS)} = 12.7 - 0.3 \cdot T - 4.8 \cdot S + 0.01 \cdot T \cdot S. \quad (4)$$

On the other hand, the equivalent phenolic content measured by the F–C method was only affected by temperature: the higher the temperature, the lower the F–C content, ranging from 5.5 mg gallic acid per g of DSL at 250°C to 1.3 mg gallic acid per g of DSL at 325°C.

3.4 FTIR analysis

As can be seen in Fig. 5, significant differences could be observed among the samples subjected to hydrothermal treatment in comparison to the original BLS. One of the most relevant observation that could be made is the notorious increase in the broad peak (3500–3050 cm^{-1}), corresponding to the phenol and alcohol O-H stretching, which are the most abundant compounds.

According to previous studies [14], [15], the peaks in the 2970–2720 cm^{-1} region are related to C-H stretching in alkenes and alkanes; the bands between 1576 and 1542 cm^{-1} are characteristic of aromatic ring vibrations. Phenolic ring and O-H are responsible of the

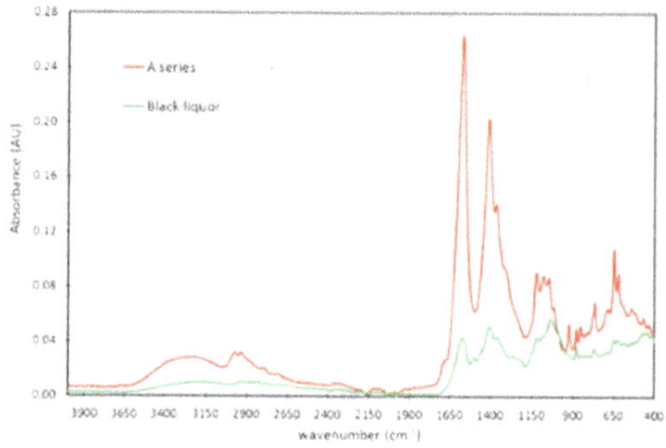

Figure 5: FTIR-ATR spectra of the hydrothermally treated samples (average of all SDL samples) compared to the original BLS.

bands ranging from 1413 to 1337 cm^{-1}, while peaks in the 1121–1088 cm^{-1} region are due to C-O stretching. Finally, the bands observed at the lowest wavenumbers (923–616 cm^{-1}) correspond to out-of-plane C-H bending. In order to get the maximum information, first derivative of spectra was obtained. Six bands were significant from a statistical point of view. The Pearson correlation coefficient was calculated and values higher than 0.850 were considered as relevant. Only a positive correlation (0.927) was found between bands at 1141 and 1610 cm^{-1}, whereas three negative correlations were noticed between bands at 1355 and 1610 cm^{-1} (-0.943), 1355 and 1446 cm^{-1} (-0.861), and 885 and 887 cm^{-1} (-0.970). In any case, only small differences were found among the FTIR spectra of hydrothermally treated samples obtained under different conditions.

3.5 Effect of the addition of depolymerized lignin fractions on the biodiesel oxidation stability

The biodiesel oxidation stability time (OXY) was measured prior and after adding the samples of SDL to biodiesel. Table 6 shows the results of the impact of adding the SDL samples to biodiesel on the values obtained from the OXY test. Although data in Table 6 seem to point that, for the same reaction time and BLS content, the biodiesel oxidation stability is higher when adding SDL produced at 325°C than when adding those obtained at 250°C, the ANOVA test does not show a clear trend with any of the experimental factors, probably because of the great variability observed in the replicates at the central point.

In order to correlate the properties of the characterized SDL with the biodiesel oxidation stability, a principal component analysis (PCA) was performed taking into account the intensity of the FTIR bands, the average Mn, the concentration of the GC/MS families, (organized by functional groups) and the Folin–Ciocalteu (F–C) phenolic concentration. The scores plot (Fig. 6(a)) shows that the first two principal components explain 90.1% of the total variance, where temperature is the most influential parameter. Data could be grouped in three clusters corresponding to the three tested temperatures. Fig. 6(b) shows the correlation loadings. Results show that the FTIR bands at 1561 and 1355 cm^{-1} are moderately related with the OXY increase (correlation 0.665 and 0.510), whereas there is a strong correlation between phenolics and ethers (by GC/MS, 0.932) and phenolics (by GC/MS) vs. F–C concentration (0.700). Phenolics show moderate correlation with Mn (0.582) and the IR band at 1446 cm^{-1} (0.571). The worst agreement (0.451) is found between phenolics and OXY, thus indicating that the improvement in the antioxidant properties is not only due to the presence of volatile phenolics, being also significant the compounds present into the non-volatile fraction.

Table 6: OXY improvement of biodiesel after adding the SDL samples.

Temp (°C)	Time (h)	BLS content (wt.%)	ΔOXY (%)
250	1	7	19.8
325	1	7	144
250	3	7	31.7
325	3	7	54.8
250	1	14	8.7
325	1	14	33.6
250	3	14	13
325	3	14	24.7
287.5	2	10.5	78.7
287.5	2	10.5	94.7
287.5	2	10.5	83.9

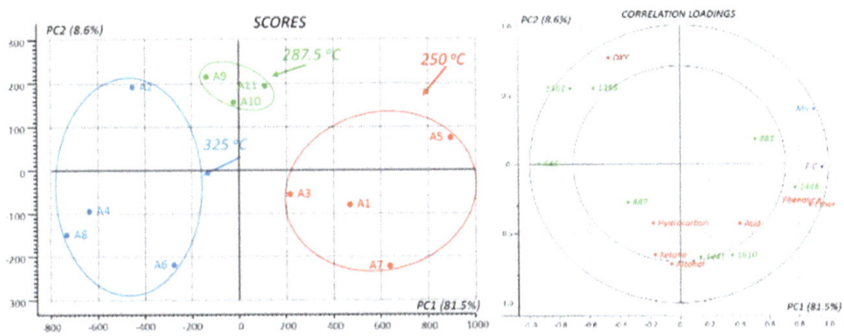

(a) (b)

Figure 6: (a) Scores; and (b) Correlation loadings plots of principal components 1 (81.5% of explained variance) and 2 (8.6%).

4 CONCLUSIONS

Black liquor from straw was produced and depolymerized in a batch reactor. The depolymerized lignin was tested as antioxidant for biodiesel. Although the observed increase on the biodiesel oxidation stability was moderate, the results show potential for the valorization of this byproduct of the pulping industry. The depolymerized lignin was characterized by GPC, GC/MS, Folin–Ciocalteu phenolic concentration and FTIR spectroscopy. The effect of the experimental conditions was noticeable on the molecular weight distribution measured by GPC, especially the effect of reaction temperature, which was overall identified as the most influential parameter. The increased temperature decreased the molecular weight, as well as the phenolic content, measured either by GC/MS or with the Folin–Ciocalteu method. FTIR showed the presence of functional groups that can be assigned to phenolic compounds, as aromatic ring, phenolic rings and OH vibrations. The oxidation stability increase could not be clearly correlated with the analyzed parameters of the depolymerized BL.

ACKNOWLEDGEMENTS

The authors acknowledge the funding from the Aragón Government (Ref. T22_17R), co-funded by FEDER 2014-2020 *"Construyendo Europa desde Aragón"* and *Ministerio de Economía y Competitividad (Gobierno de España)* for the project ENE2017-85040-R: Renewable additives for biofuels.

REFERENCES

[1] Lora, J.H. & Escudero, E., Soda pulping of agricultural fibres for boardmaking applications. *Paper Technol.*, **41**, pp. 37–42, 2000.

[2] Ono, K., Isono, Z. & Tsuda, T., Desilicification from the high silica containing kraft black liquor by the improved carbon dioxide method. *Agric. Biol. Chem.*, **33**, pp. 622–624, 1969.

[3] Marín, F. et al., Semichemical pulping of Miscanthus giganteus. Effect of pulping conditions on some pulp and paper properties. *Bioresour. Technol.*, **100**, pp. 3933–3940, 2009.

[4] Gonzalo, A. et al., Evaluation of different agricultural residues as raw materials for pulp and paper production using a semichemical process. *J. Clean. Prod.*, **156**, pp. 184–193, 2017.

[5] Mahmood, N. et al., Hydrolytic depolymerization of hydrolysis lignin: Effects of catalysts and solvents. *Bioresour. Technol.*, 2015. DOI: 10.1016/j.biortech.2015.04.074.

[6] Beauchet, R., Monteil-Rivera, F. & Lavoie, J.M., Conversion of lignin to aromatic-based chemicals (L-chems) and biofuels (L-fuels). *Bioresou.r Technol.*, **121**, pp. 328–334, 2012.

[7] García, M. et al., Antioxidants for biodiesel: Additives prepared from extracted fractions of bio-oil. *Fuel Process Technol.*, **156**, pp. 407–414, 2017.

[8] Lange, H., Rulli, F. & Crestini, C., Gel permeation chromatography in determining molecular weights of lignins: Critical aspects revisited for improved utility in the development of novel materials. *ACS Sustain. Chem. Eng.*, **4**, pp. 5167–5180, 2016.

[9] Asikkala, J. & Tamminen, T., Argyropoulos DS: Accurate and reproducible determination of lignin molar mass by acetobromination. *J. Agric. Food Chem.*, **60**, pp. 8968–8973, 2012.

[10] Ristolainen, M., Alén, R. & Knuutinen, J., Characterization of TCF effluents from kraft pulp bleaching. I. Fractionation of hardwood lignin-derived material by GPC and UF. *Holzforschung*, **50**, pp. 91–96, 1996.

[11] Pakkanen, H. & Alén, R., Molecular mass distribution of lignin from the alkaline pulping of hardwood, softwood, and wheat straw. *J. Wood Chem. Technol.*, **32**, pp. 279–293, 2012.

[12] Kissin, Y.V., Molecular weight distributions of linear polymers: Detailed analysis from GPC data. *J. Polym. Sci. Part A: Polym. Chem.*, **33**, pp. 227–237, 1995.

[13] Clementi, L.A. et al., Molar mass distributions in homopolymer blends from multimodal chromatograms obtained by Sec/Gpc with a concentration detector. *Polym. Test*, **43**, pp. 58–67, 2015.

[14] El Mansouri, N.-E. & Salvadó, J., Analytical methods for determining functional groups in various technical lignins. *Ind. Crops Prods.*, **26**, pp. 116–124, 2007.

[15] Derkacheva, O. & Sukhov, D., Investigation of lignins by FTIR spectroscopy. *Macromol. Symp.*, **265**, pp. 61–68, 2008.

ENERGY AND SUSTAINABILITY ASSESSMENT OF MUNICIPAL WASTEWATER TREATMENT UNDER CIRCULAR ECONOMY PARADIGM

ANDREY V. KISELEV[1,2], ELENA R. MAGARIL[2] & ELENA CRISTINA RADA[3]
[1]Department of Investment Program, Municipal Unitary Enterprise for Water Supply and Sewerage, Russia
[2]Department of Environmental Economics, Ural Federal University, Russia
[3]Department of Theoretical and Applied Science, Insubria University, Italy

ABSTRACT

Climate change and anthropogenic pollution have put limited water resources under pressure. Lack of basic sanitation services as well as the discharge of improperly treated effluent from wastewater treatment plants (WWTP) result in the deposition of large amounts of organic matter and nutrients, which have major detrimental effects on health. Wastewater treatment (WWT) can reduce water pollution but at the cost of increasing energy consumption and the corresponding atmosphere and climate problems. Sustainable WWT management is a global challenge to preserve fresh water and decrease energy consumption. Nowadays it becomes obvious that existing WWTP operation model, based on the linear "take–make–dispose" pattern, is no longer sustainable. Furthermore, disposal of a product in landfill means that all residual energy is lost. The adoption of circular economy (CE) practices with its 3R principles of reducing, reusing and recycling material appears as a timely, relevant and practical option to meet the goals of sustainable development. WWTP is a critical element in CE implementation policy and to measure the degree of "circularity" there is a need for indicators. This study considers the holistic overview of measuring the progress of CE implementation at WWTP under 3R principles using life cycle analysis (LCA) and material flow analysis (MFA) frameworks. The paper presents the principles of CE indicators set construction using managerial approach. The proposed set of indicators and integral circularity index are studied under three scenarios, based on real performance of northern and southern WWTP in Ekaterinburg, Russia. This study provides an efficient assessment tool of CE progress, which is rather simple for calculation and interpretation and suitable for the use of wide range of stakeholders.

Keywords: circular economy, wastewater treatment, sustainable management, energy flow, material flow, assessment, circularity index.

1 INTRODUCTION

Water scarcity is one of the main problems faced by many societies in the 21st century and will become one of the most sensitive environmental issues in the coming decades [1]. At the current time, more than 2 billion people are living with the risk of reduced access to freshwater resources and by 2050, at least one in four people is likely to live in a country affected by chronic or recurring shortages of fresh water [2].

Climate change and anthropogenic pollution have put limited water resources under pressure. Lack of basic sanitation services as well as the discharge of improperly treated effluent from wastewater treatment plants (WWTP) result in the deposition of large amounts of organic matter and nutrients, which have major detrimental effects on health [3]. As the result, eutrophication of surface water and coastal zones is expected to increase almost everywhere until 2030 [4]. Wastewater treatment (WWT) can reduce water pollution but at the cost of increasing energy consumption and the corresponding atmosphere and climate problems [5], [6]. Different solution for wastewater management are present in over the word demonstrating the necessity to find a sustainable and friendly solution from environmental, economic and energy point of view [7]–[9].

WIT Transactions on Ecology and the Environment, Vol 237, © 2019 WIT Press
www.witpress.com, ISSN 1743-3541 (on-line)
doi:10.2495/ESUS190101

Sustainable WWT management is a global challenge to preserve fresh water and decrease energy consumption. According to UNESCO [10], at present only 20% of globally produced wastewater receives proper treatment with generation about 140 million tons of dry matter sewage sludge. The last figure was counted, using EurObserv'ER Report [11] with approximation on world's population. WWT process requires significant amounts of energy ranging from 250 to 500 kWh/million L of treated wastewater with the growth perspective by 44% up to 2030, especially for non-OECD countries, where wastewater currently receives little or no treatment [10].

Nowadays it becomes obvious that existing WWTP operation model, based on the linear "take–make–dispose" pattern, in no longer sustainable. Furthermore, disposal of a product in landfill means that all residual energy is lost [12], [13]. Now, the adoption of circular economy (CE) practices appears as a timely, relevant and practical option to meet the goals of sustainable development [14].

CE is a sustainable development strategy that is being proposed to tackle urgent problems of environmental degradation and resource scarcity. CE with its 3R principles of reducing, reusing and recycling material clearly illustrates the strong linkages between the environment and economics [15]. A sustainable system is characterized, among others, by a much reduced use of renewable and non-renewable inputs and closed loop reuse and recycling of material outputs, thus drastically reducing or eliminating waste and dissipative loss [16].

WWTPs possess a large potential to improve WWT in order to reduce insufficiently treated effluent and recover wastewater resources such as nutrients, energy, and the water itself [17].

One of the critical questions in the CE policy implementation is how we should measure its performance, since its objectives are substantially different from those in the traditional linear economy [18]. The core idea of measuring "circularity" is to see how materials are enter, flow within and (eventually) leave the economy [19]. The monitoring framework aimed at assessment the progress of CE implementation throughout different dimensions at all stages of lifecycle resources, products and services. To assess progress towards a more CE and the effectiveness of action, it is important to have a set of reliable indicators [20]. WWTP is a critical element in CE implementation policy and to measure the degree of "circularity" there is a need for a set of indicators.

The analysis of literature shows that most of existing studies are devoted to CE monitoring framework based on reuse and recycle applications. CE is not only reuse or recycle applications, but also reduce practices implementation, especially relevant for developing countries. That is why 3R principles should be taken into account within CE monitoring framework of WWTP. Moreover, proposed indicators are complex for calculations and require access to large amount of input data.

The more complex and intensive WWT process, the greater energy consumption, thus the aim of current research is to create the monitoring framework of WWTP efficiency under CE principles in order to ensure the proper quality of discharged wastewater with moderate energy consumption.This study considers the holistic overview of measuring the progress of CE implementation at WWTP under 3R principles using life cycle analysis (LCA) and material flow analysis (MFA) frameworks. Flows of material and energy are the core of circular economy and hence circular economy indicators should be able to capture these dimensions [12], [13], [16].

2 MATERIALS AND METHODS

The developed by authors method includes the following steps: (i) study area description; (ii) monitoring framework design using LCA and MFA under 3R principle; (iii) CE indicators set for WWTP definition; and (iv) integral circularity index development. A managerial approach is used for results interpretation, ranging output values for all CE indicators from 0 (which mean perfect linearity) to 1 (which mean perfect circularity).

2.1 Investigated scenarios definition

Three scenarios for WWTP operation systems are considered within this paper (Fig. 1) based mainly on the centralized sewerage system of Ekaterinburg, Russia. Ekaterinburg is the fourth largest city in the Russian Federation with the population of 1,468,833 inhabitants [21]. It is the administrative centre of the Sverdlovsk region, located on the border of Europe and Asia, the largest industrial, educational, scientific, touristic, commercial and financial center, as well as a transport and logistics hub.

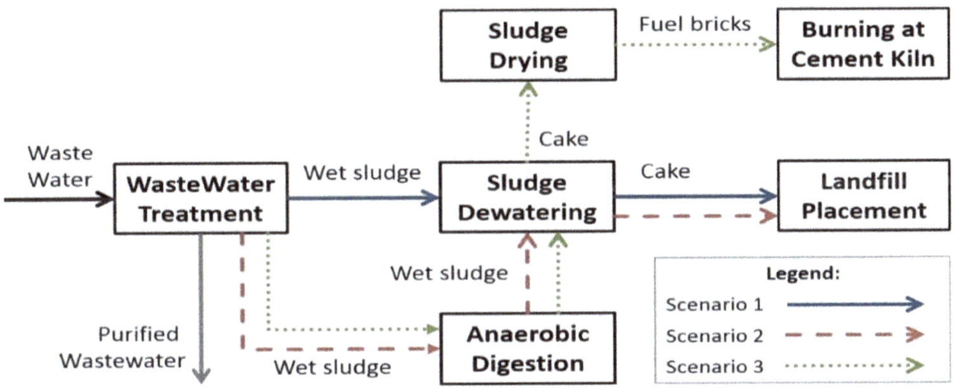

Figure 1: Investigated scenarios definition. (a) Scenario 1, based on southern WWTP in Ekaterinburg, Russia; (b) Scenario 2, based on northern WWTP in Ekaterinburg, Russia; (c) Scenario 3, based on best available techniques.

The Ekaterinburg sewerage system has been built on the basin principle: two main sewerage zones can be identified within the city – northern and southern ones. WWT from these zones is carried out at the northern and southern WWTP, respectively. The southern WWTP was designed in 1970 with maximum wastewater inflow up to 550,000 m^3 per day, while the northern one has maximum performance up to 100,000 m^3 per day.

The first scenario shown at Fig. 1 is based on the WWT process actually performed at southern WWTP. WWT includes primary (mechanical) and secondary (biological) treatment with chlorine disinfection before discharge. Sewage sludge treatment include mechanical dewatering with cake placement at landfills. This WWTP operates without significant modernization since its foundation and has obsolete technologies.

The second scenario shown at Fig. 1 is based on the WWT process actually performed at northern WWTP. The basic WWT operations are similar to southern WWTP, however in 2018 total modernization of the process line was finished, introducing best available techniques (BAT) including rotary drum fine screens, sand traps with aeration, aeration tanks with nitrification and denitrification, UV-disinfection before discharge and others. Sewage sludge treatment include anaerobic digestion with biogas generation and mechanical dewatering. Biogas CHP-unit installation is planned in nearest future.

The third scenario shown in Fig. 1 is based on the second scenario supplemented with company's future modernization plans, including sewage sludge drying with incineration at cement kiln.

2.2 Methodology framework for construction of CE indicators

This study was performed using LCA and MFA framework as a basic model for CE indicators construction under 3R principles of CE.

LCA is related to the ISO 14000 environmental management standard. LCA is the assessment of the environmental impact of a given product throughout its lifespan. The goal of LCA is to compare the environmental performance of products in order to be able to choose the least burdensome [22]. All activities and processes result in environmental impacts due to consumption of resources, emissions of substances into the natural environment, and other environmental exchanges [23]. LCA technique has a step-by-step practice according to ISO 14040:2006, including four stages: (i) goal and scope definition; (ii) inventory analysis; (iii) impact analysis; and (iv) interpretation of results.

The approach with most direct relevance for CE indicators is MFA, based on the principles and methods of mass balancing and input–output analysis [16]. MFA is an accounting system that captures the mass balances in an economy, where inputs are equal to outputs, and based on fundamental principles, specifically the laws of thermodynamics [24]. The core idea of MFA is to specify and quantify the pathways of materials/energy into, through and out of specified system boundaries [25]. It is expressed in material flows monitoring using physical units (weight) and its qualitative rating.

The purpose of this study is not the use of the LCA and MFA techniques in step-by-step actions within selected scenarios, but the use of these tools as a framework in construction of the CE indicators. In this way, WWTP LCA framework, developed by Buonocore et al. [23], was taken as a basic model for further work. This framework has undergone some improvements with MFA elements and matches all scenario under consideration (Fig. 2).

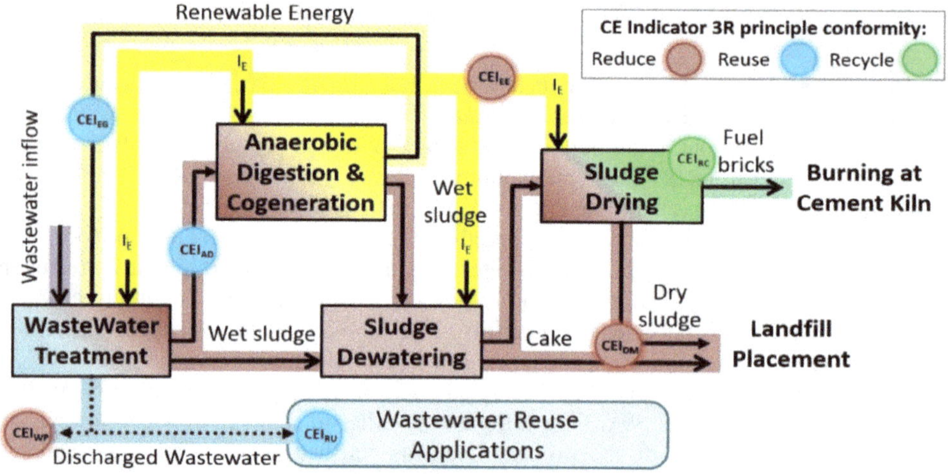

Figure 2: CE indicators set under 3R principle. Blue line = wastewater flow; red line = sewage sludge flow; yellow line = energy flow; green line = recycled sewage sludge flow.

Main streams of materials were determined including wastewater as an inflow, purified wastewater and sludge as outflow. It is worth noting the separate flow of recycled sludge as the possible option for WWTP. Energy flow is also considered within selected framework, including energy input from grid and renewable energy from anaerobic digestion. CE indicators are derived from considered framework (Fig. 2) as outflow checkpoints, which corresponds to the transition towards CE within 3R categories.

Analysis of review papers allowed to construct CE indicator set, which implies single CE indicator for each 3R category within any flow, except recycle – in this category only sewage sludge is possible to recycle. As the result, seven CE indicators make up a set for energy and sustainability assessment of municipal WWT under CE paradigm. This set include:

1. Reduce CE indicator for wastewater flow (CEI_{WP}).
2. Reduce CE indicator for sewage sludge flow (CEI_{DM}).
3. Reduce CE indicator for energy flow (CEI_{EE}).
4. Reuse CE indicator for wastewater flow (CEI_{RU}).
5. Reuse CE indicator for sewage sludge flow (CEI_{AD}).
6. Reuse CE indicator for energy flow (CEI_{EG}).
7. Recycle CE indicator for sewage sludge flow (CEI_{RC}).

2.2.1 Reduce CE indicator for wastewater flow

CEI_{WP} measures the efficiency of WWT. The basic principle was mentioned by Rukavishnikova et al. [26] within the WWT quality assessment approach using multiplicity of wastewater samples with some improvements by the authors. Multiplicity (M) is the intensity of negative exposure of harmful substances (pollutants) on the water body [22]. Multiplicity of pollutant i is calculated as follows:

$$M_i = \frac{C_i^P}{MPC_i^P}, \tag{1}$$

where C_i^P = annual average concentration of i substance and MPC_i^P = maximum permissible concentration of i substance. The most critical pollutants for WWTP BAT technology conformity were identified by the authors, including: (i) suspended solids; (ii) biochemical oxygen demand in 20 days (BOD_{20}); (iii) phosphorus phosphate; (iv) nitrate-ion; (v) nitrite-ion; and (vi) nitrogen ammonia. CEI_{WP} is calculated as follows:

$$CEI_{WP} = \frac{\sum_{i=1}^{6} MW_i}{6}, \tag{2}$$

where MW_i = multiplicity weight of each substance.

Multiplicity weight is an expert evaluation of multiplicity in relation to the CE. It means that the more pollutants above the allowed MPC enters the water bodies, the lower the CE indicator value. Multiplicity weight values are presented in Table 1.

Table 1: Multiplicity weight values.

	Multiplicity value (M)			
	$0 \leq M \leq 1$	$1 < M \leq 5$	$5 < M \leq 15$	$M > 15.01$
Multiplicity weight (MW_i)	1.00	0.50	0.25	0.00

Further investigations can set additional rates for multiplicity weight values considering the difference in hazard degree of substances.

2.2.2 Reduce CE indicator for sewage sludge flow

The volume of output sewage sludge depends on the dry matter (DM) content and applied dewatering techniques. Water removal is the primary solution of volume reduction [27]: the more DM content in sewage sludge – the less WWTP personnel is required to export it to the landfills. Thus, the progress of CE is directly proportional to DM content of sewage sludge ready for export to landfills. CEI_{DM} is calculated as follows:

$$CEI_{DM} = \text{DM} = \frac{M_{SS}^{DM}}{M_{SS}}, \tag{3}$$

where, M_{SS}^{DM} = total mass of dry mater in sewage sludge, and M_{SS} = total mass of sewage sludge. In some cases, e.g. when the sewage sludge is subjected to drying and subsequent use as a biofuel for incineration at cement kilns (in other words, it is recycled), placement on landfills is not performed, thus $CEI_{DM} = 1$ (waste is completely transformed into resource).

2.2.3 Reduce CE indicator for energy flow

CEI_{EE} measures the efficiency of WWTP's energy consumption. CEI_{EE} is selected according to net energy efficiency weight values (Table 2), designed by authors, using specific net energy efficiency indicator (EE_{NET}), which is calculated as follows:

$$EE_{NET} = \frac{EC_T - EG_R}{Q_{TWW}}, \tag{4}$$

where, EC_T = total WWTP energy consumption, related to the full WWT cycle, including sludge treatment, EG_R = total WWTP self-generated energy from renewable sources (e.g. biogas), Q_{TWW} = total volumetric flow of treated wastewater. Net energy efficiency weight is an expert evaluation of net energy efficiency in relation to the CE (see Table 2). This interpretation was set up according to the typical WWTP energy consumption, mentioned by Electric Power Research Institute and Water Research Foundation [28] and Gandiglio et al. [29] and is given for two different WWTP scales.

Table 2: Net energy efficiency weight values.

#	EE_{NET} Value (kWh/m^3)	CEI_{EE}	CEI_{EE}
		V ≥ 30,000 m^3/day	V ≤ 30,000 m^3/day
1	$EE_{NET} \leq 0.300$	1.00	1.00
2	$0.300 < EE_{NET} \leq 0.440$	0.80	1.00
3	$0.440 < EE_{NET} \leq 0.700$	0.60	0.80
4	$0.700 < EE_{NET} \leq 1.000$	0.40	0.60
5	$1.000 < EE_{NET} \leq 1.500$	0.20	0.40
6	$1.500 < EE_{NET} \leq 1.800$	0.00	0.20
7	$EE_{NET} > 1.800$	0.00	0.00

2.2.4 Reuse CE indicator for wastewater flow

CEI_{RU} measures the reuse of already treated wastewater. Similar indicator was mentioned by Molina-Sánchez et al. [30], Geng et al. [24] and is defined as the treated wastewater flow that can be obtained and reused by alternative users, for example for irrigation or communal (e.g. streets washing) purposes. CEI_{RU} is calculated as follows:

$$CEI_{RU} = \frac{Q_{RU}}{Q_{TWW}}, \tag{5}$$

where Q_{RU} = volumetric flow of reused wastewater, Q_{TWW} = total volumetric flow of treated wastewater.

2.2.5 Reuse CE indicator for sewage sludge flow
Nowadays sewage sludge anaerobic digestion is considered to be state-of-the-art technique for the efficient sewage sludge management. CEI_{AD} is calculated as follows:

$$CEI_{AD} = \frac{Q_{AD}}{Q_{TSS}}, \tag{6}$$

where Q_{AD} = volumetric flow of sewage sludge that has been processed stabilization through anaerobic digestion, Q_{TSS} = total volumetric flow of generated sewage sludge.

2.2.6 Reuse CE indicator for energy flow
CEI_{EG} is closely related to CEI_{AD} and has directly proportional relationship. Pinter [16] mentioned similar indicators in measuring the progress towards "sound material cycle" and included it in his indicator set for material and energy flows. CEI_{EG} measures the share of generated renewable energy in total WWTP energy consumption and is calculated as follows:

$$CEI_{EG} = \frac{EG_R}{EC_T}, \tag{7}$$

where, EG_R = total WWTP self-generated energy from renewable sources (e.g. biogas), EC_T = total WWTP energy consumption. There is a possibility that the rate of energy generation exceeds the rate of energy consumption, however we are considering indicators at micro (company's) level and the main "circularity" issue is to cover own energy consumption.

2.2.7 Recycle CE indicator for sewage sludge flow
CEI_{RC} is the only one indicator in the recycle category. CEI_{RC} measures the recycle rate of dewatered sewage sludge (or cake). For instance, recycling can be fulfilled through sludge drying and producing organic fuel bricks for further use in cement kiln for the manufacturing of Portland cement. CEI_{RC} is calculated as follows:

$$CEI_{RC} = \frac{M_{TDSS}^{Rcl}}{M_{TDSS}}, \tag{8}$$

where, M_{TDSS}^{Rcl} = total mass of recycled dewatered sewage sludge, M_{TDSS} = total mass of dewatered sewage sludge.

2.3 Integral circularity index

Integral circularity index ($ICIn$) is intended to integrate the results of CE indicators set (mentioned in Sections 2.2.1–2.2.7). This indicator should reflect the progress towards CE without distortion at the similar output values range. Each of three 3R categories has the same importance for CE progress as well as three types of flow. Thus, $ICIn$ is calculated as follows:

$$ICIn = \frac{\sum_{i=1}^{7} CEI_i}{7}. \tag{9}$$

3 RESULTS AND DISCUSSIONS

The period from 2014 to 2018 was chosen to perform the proposed method for WWTP CE implementation progress assessment. Scenarios 1 and 2 have calculated values throughout the period mentioned above, while Scenario 3 is based on Scenario 2 with some adjustments (under modernization plans) and focused on the future perspective, starting from 2018. In other words, Scenario 3 has two time points: "2018" (which is equal to Scenario 2 CE Indicator values) and "future perspective".

Table 3 shows the result of an experiment conducted under considered scenarios using data from southern and northern WWTPs, situated in Ekaterinburg (Russia).

Table 3: CE indicators set and integral circularity index for Ekaterinburg WWTP.

#	Scenario	Year	CE Indicators							
			CEI_{WP}	CEI_{DM}	CEI_{EE}	CEI_{RU}	CEI_{AD}	CEI_{EG}	CEI_{RC}	$ICIn$
1	1	2014	0.50	0.25	0.80	0.00	0.00	0.00	0.00	0.22
2		2015	0.50	0.25	0.80	0.00	0.00	0.00	0.00	0.22
3		2016	0.50	0.25	0.80	0.00	0.00	0.00	0.00	0.22
4		2017	0.42	0.25	0.60	0.00	0.00	0.00	0.00	0.18
5		2018	0.46	0.25	0.60	0.00	0.00	0.00	0.00	0.19
6	2	2014	0.75	0.17	0.80	0.00	0.00	0.00	0.00	0.25
7		2015	0.71	0.17	1.00	0.00	0.00	0.00	0.00	0.27
8		2016	0.75	0.17	0.80	0.00	0.00	0.00	0.00	0.25
9		2017	0.58	0.25	0.80	0.00	0.00	0.00	0.00	0.23
10		2018	0.67	0.25	0.80	0.00	1.00	0.00	0.00	0.39
11	3	2018	0.67	0.25	0.80	0.00	1.00	0.00	0.00	0.39
12		F.p.*	0.67	1.00	1.00	0.00	1.00	0.67	1.00	0.76

*Future perspective.

CE is not only reuse and recycle applications, but also reduce ones, especially popular in developing countries. Thus, the use of 3R principle within the set of indicators is of a great importance. According to survey results, the first three CE indicators from reduce category have more sound values in comparison with reuse and reduce categories. These values make up the main part of integral circularity index. To some extent, reduce applications can be considered the first step towards CE. Reduction refers to minimize the input of primary energy and raw materials through the improvement of production efficiency [31].

Reuse of goods or resources suggests the use of a product again for the same purpose in its original form or with little enhancement or change [12]. Wastewater reclamation is considered one of the recommended solutions for the problem of water scarcity. Some countries, including Cyprus, France, Greece, Spain, Italy and Portugal, have standards developed specially for the reuse of treated wastewater [1]. Recycle encourages processing the recyclable materials into new products so that the consumption of virgin materials can be reduced. These principles, as parts of the whole process, have different hierarchical importance, with the reduction of resources used as the leading principle within a circular economy system [31].

The progress towards CE for scenarios under consideration starting from 2014 in terms of integral circularity index is presented in Fig. 3.

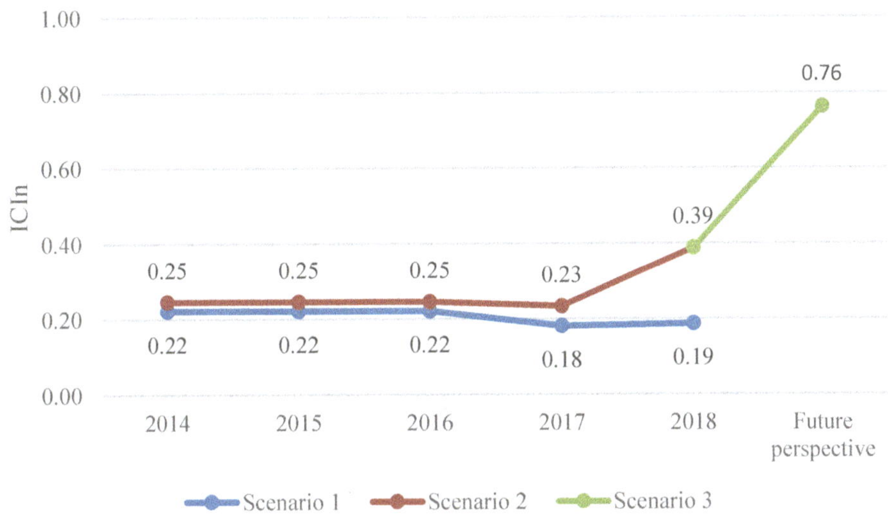

Figure 3: The progress towards CE for Ekaterinburg WWTPs.

The results show that current investment strategy for Ekaterinburg WWTP inevitably leads towards CE. The uprising trend of $ICIn$ for scenario 2 refers to successful implementation of modernization project for northern WWTP in 2018 with full WWT technology upgrade. Next step towards CE for northern WWTP includes CHP-unit installation and the construction of sewage sludge drying plant, which is referred to scenario 3. In this way, northern WWTP can achieve 76% of circularity. On the other hand, southern WWTP is going towards linearity and there is a strong need for deep modernization using proven solutions that are already applied at northern WWTP.

The reuse of effluent (e.g. for irrigation) in Russian Federation and some other countries is limited by legal or sanitary restrictions or stumble over cultural issues (some people are disgusted if they find out that crops are produced with treated domestic wastewater) thus the achievement of high circularity rates is difficult. However, there is an alternative to use effluent for industrial and communal purposes, e.g. streets cleaning, cooling for machines and mechanisms, centralized heating, construction of objects etc.

4 CONCLUSION

The proposed set of indicators and integral circularity index create an efficient assessment tool, which is rather simple for calculation and interpretation and suitable for the use of wide range of stakeholders. In comparison with reviewed papers this tool does not require significant input data and is easy to understand to what extent the WWTP is circular or linear.

It is important tool for governments and business to measure the progress towards CE. Moreover, integral circularity index can be used as a benchmark for WWTP sector within region, country or worldwide to promote continuous improvement and best practice exchange.

However, there are some limitations for applying this approach. CE indicators set is based on the principle of measuring the average values of each flow (e.g. volumetric units) and doesn't take into account the scale of WWTP. There are possible options when integrated circularity index is close to maximum values, but the component of WWT quality is moderate and within a large scale WWTP a huge amount of waste load enters the water bodies. Further researches with different CE indicator weights under multiscale framework for Integral circularity index construction are expected.

ACKNOWLEDGEMENT

This research was supported by Act 211 Government of the Russian Federation, contract № 02.A03.21.0006.

REFERENCES

[1] Pintilie, L., Torres, C.M., Teodosiu, C. & Castells, F., Urban wastewater reclamation for industrial reuse: An LCA case study. *Journal of Cleaner Production*, **139**, pp. 1–14, 2016. DOI: 10.1016/j.jclepro.2016.07.209.

[2] United Nations Sustainable Development, www.un.org/sustainabledevelopment. Accessed on: 29 Apr. 2019.

[3] Naidoo, S. & Olaniran, A.O., Treated wastewater effluent as a source of microbial pollution of surface water resources. *Int. J. Environ. Res. Public Health*, **11**, pp. 249–270, 2014. DOI:10.3390/ijerph110100249.

[4] United Nations Department of Economic and Social Affairs, Back to our common future: Sustainable development in the 21st century (SD21) project, 2012. https://sustainabledevelopment.un.org/content/documents/UN-DESA_Back_Common_Future_En.pdf. Accessed on: 29 Apr. 2019.

[5] Niu, K., Wu, J., Qi, L. & Niu, Q., Energy intensity of wastewater treatment plants and influencing factors in China. *Science of the Total Environment*, **670**, pp. 961–970, 2019. DOI: 10.1016/j.scitotenv.2019.03.159.

[6] Pajares, E.M., Valero, L.G. & Sánchez, I.M.R., Cost of urban wastewater treatment and ecotaxes: Evidence from municipalities in southern Europe. *Water,* **11**(3), p. 423, 2019.

[7] Langone, M. et al., Application of a novel hydrodynamic cavitation system in wastewater treatment plants. *UPB Scientific Bulletin, Series D*, **77**(1), pp. 225–234, 2015.

[8] Ragazzi, M., Catellani, R., Rada, E.C., Torretta, V. & Salazar-Valenzuela, X., Management of urban wastewater on one of the Galapagos Islands. *Sustainability*, **8**(3), p. 208, 2016.

[9] Eggimann, S., Truffer, B., Feldmann, U. & Maurer, M., Screening European market potentials for small modular wastewater treatment systems: An inroad to sustainability transitions in urban water management? *Land Use Policy*, **78,** pp. 711–725, 2018.

[10] UNESCO, Managing water under uncertainty and risk, the UN world water development report 4, vol. 1, 2012. http://unesdoc.unesco.org/images/0021/002156/215644e.pdf. Accessed on: 29 Apr. 2019.

[11] EurObserv'ER, The state of renewable energies in Europe. Edition 2016. www.eurobserv-er.org/pdf/2016/EurObservER-Annual-Overview-2016-EN.pdf. Accessed on: 12 Mar. 2019.

[12] Ellen MacArthur Foundation, Towards the circular economy: economic and business rationale for an accelerated transition, 2013.

[13] Ellen MacArthur Foundation, Circularity indicators: An approach to measuring circularity. Methodology, 2015. www.ellenmacarthurfoundation.org/assets/downloads/insight/Circularity-Indicators_Methodology_May2015.pdf. Accessed on: 29 Apr. 2019.
[14] Saidani, M., Yannou, B., Leroy, Y., Cluzel, F. & Kendall, A., A taxonomy of circular economy indicators. *Journal of Cleaner Production*, **207**, pp. 542–559, 2019.
[15] Heshmati, A., A review of the circular economy and its implementation. IZA Discussion Paper, No. 9611, 2015.
[16] Pinter, L., International experience in establishing indicators for the circular economy and considerations for China. Report for the Environment and Social Development Sector Unit, East Asia and Pacific Region, The World Bank, 2006.
[17] van Leeuwen, K., de Vries, E., Koop, S. & Roest, K., The energy and raw materials factory: role and potential contribution to the circular economy of the Netherlands. *Environmental Management*, **61**, pp. 786–795, 2018. DOI: 10.1007/s00267-018-0995-8.
[18] EASAC, Policy Report 2016, Indicators for circular economy. www.easac.eu/fileadmin/PDF_s/reports_statements/Circular_Economy/EASAC_Indicators_web_complete.pdf. Accessed on: 29 Apr. 2019.
[19] Communication from the commission to the European Parliament, the Council, the European economic and social committee and the committee of the regions on a monitoring framework for the circular economy, Strasbourg, 2018.
[20] EC-European Commission Closing the Loop – an EU Action Plan for the Circular Economy, The Circular Economy Package Proposal, Brussels, Belgium, 2015.
[21] Wikipedia, Population of Ekaterinburg. (In Russian.) https://ru.wikipedia.org/wiki/Население_Екатеринбурга. Accessed on: 29 Apr. 2019.
[22] ISO 14040:2006, Environmental management: Life cycle assessment. www.iso.org/standard/37456.html. Accessed on: 29 Apr. 2019.
[23] Buonocore, E., Mellino, S., De Angelis, G., Liu, G. & Ulgiati, S., Life cycle assessment indicators of urban wastewater and sewage sludge treatment. *Ecological Indicators*, **94**, pp. 13–23, 2018. DOI: 10.1016/j.ecolind.2016.04.047.
[24] Geng, Y., Fu, J., Sarkis, J. & Xue, B., Towards a national circular economy indicator system in China: An evaluation and critical analysis. *Journal of Cleaner Production*, **23**, pp. 216–224, 2012.
[25] Cai, Z.C. & Wennerstern, R., Ecological urban design through material and energy flow analysis and life cycle assessment: From an architect's perspective. *WIT Transactions on Ecology and the Environment*, vol. 142, pp. 3–13, 2010.
[26] Rukavishnikova, I., Kiselev, A., Berezyuk, M. & Ashirova, I., Improvement of the methodology for assessing domestic wastewater treatment quality using benchmarking tools. *WIT Transactions on Ecology and the Environment*, vol. 228, pp. 209–219, 2018.
[27] Kiselev, A., Magaril, E., Magaril, R., Panepinto, D., Ravina, M. & Zanetti, M.C., Towards circular economy: Evaluation of sewage sludge biogas solutions. *Resources*, **8**(2), p. 91, 2019. DOI: 10.3390/resources8020091.
[28] Electric Power Research Institute and Water Research Foundation, Electricity use and management in the municipal water supply and wastewater industries, 2013.
[29] Gandiglio, M., Lanzini, A., Soto, A., Leone P. & Santarelli M., Enhancing the energy efficiency of wastewater treatment plants through co-digestion and fuel cell systems. *Frontiers in Environmental Science*, **5**(70), 2017.

[30] Molina-Sánchez, E., Leyva-Díaz, J.C., Cortés-García, F.J. & Molina-Moreno, V., Proposal of sustainability indicators for the waste management from the paper industry within the circular economy model. *Water*, **10**(1014), 2018. DOI: 10.3390/w10081014.
[31] Su, B., Heshmati, A., Geng, Y. & Yu, X., A review of the circular economy in China: Moving from rhetoric to implementation. *Journal of Cleaner Production*, **42**, pp. 215–227, 2013. DOI: 10.1016/j.jclepro.2012.11.020.

STUDENTS' ATTITUDES TOWARDS NUCLEAR ENERGY: RUSSIAN AND ITALIAN EXPERIENCE

ANZHELIKA KARAEVA[1], ELENA R. MAGARIL[1] & ELENA CRISTINA RADA[2,3]
[1]Department of Environmental Economics, Ural Federal University, Russian Federation
[2]Department of Civil, Environmental and Mechanical Engineering, University of Trento, Italy
[3]Department of Theoretical and Applied Science, University of Insubria, Italy

ABSTRACT

Nowadays, the energy sector faces a lot of challenges because of environmental issues, the potential depletion of fossil fuels in the future, problems of technological development, etc. Despite the fact that alternative energy is a key factor of the energy sector's development, it still has several controversial questions, and one of them is "to what extent is nuclear energy safe for the world?" As a consequence, public acceptability has a role to play as it significantly affects the way nuclear energy should be developed in the next decades. In this paper, we provide the main results of the research made in December 2018 in Italy and Russia on students' attitude towards nuclear energy and its development in their homelands. The research method is a questionnaire which was carried out at the University of Trento, Italy, and at Ural Federal University, Russia. According to the given results, most of the students from both countries do not have a negative perception of nuclear energy. However, the opinions on its development in Italy and Russia are completely different: Russian students support the government's energy program in which nuclear energy is one of the most promising trends for the future, whereas Italian students do not accept the possibility that nuclear energy could come back into the Italian energy sector.

Keywords: nuclear energy, public acceptability, environment, green energy.

1 INTRODUCTION

Currently, the energy sector is undergoing significant changes: more and more countries are striving to switch to "green" energy, thereby reducing the level of anthropogenic impact on the environment. At the same time, circular economy is giving the energy sector a different role [1], [2]. In a number of countries, nuclear energy is characterized as the energy of the future and referred to as "green". According to the European Environmental Energy report [3], this area of energy has great prospects and may become a key factor in resolving the fuel and energy crisis in the future. The fundamentals of the nuclear energy generation are well-known, and pros and cons of its application are widely discussed [4]–[7]. The nuclear power generation demonstrates the highest level of environmental characteristics and the nuclear power plants give a high power output [8]. Moreover, the final cost of generated energy is much cheaper than energy produced by using fossil fuels. However, the lessons of accidents in the past, such as Chernobyl and Fukushima, moved the issues of safety and security at the nuclear power plants to the forefront, which contributed to the rapid development of new safety technologies [9] and raised requirements for personnel skills [10].

According to the concept of sustainable development, it is necessary that nuclear power meets a number of criteria, which can generally be divided into political, social, environmental and welfare [11]. The group of social criteria can be called one of the most controversial and constraining the development of the industry in question. Due to the negative attitude of the population and the potential risks in a number of states, there are significant difficulties in its development [12]. For example, the governments of most developed countries, including Germany, Italy, Austria, the Czech Republic, and others, have suspended or completely abandoned the use of nuclear energy for peaceful purposes. The

www.witpress.com, ISSN 1743-3541 (on-line)
doi:10.2495/ESUS190111

main reason for failure is the potential risks that can cause irreparable harm to humanity and the environment and the negative attitude of the population [13].

Public opinion, in this case, directly correlates with the attitude of the government on the use of nuclear energy. More than 40% of US (which is one of the leaders of nuclear energy in the world) residents support the construction of new nuclear power plants (NPP) even after the Fukushima accident that occurred in 2013 [14], while less than 7% of German citizens showed a positive attitude towards the resumption of their operation and the start of construction of new NPP [15]. It should be noted that Germany decommissioned all nuclear reactors immediately after the Fukushima accident in 2013.

Public opinion is a serious development barrier that can significantly restrain the development of technology and energy in general. The study [16] also revealed several components from which public opinion can vary significantly within the framework of a given topic: the level of awareness of the main aspects of the use of a particular type of energy, its perception by the population, the presence of fears and age.

In addition, there is a construal level theory (CLT), which is the most relevant nowadays when considering issues related to public opinion and nuclear energy [17]. In the framework of CLT, it is assumed that the greater the distance between the subject of perception and man, the more abstract the image is formed in human thinking [18]. So, for example, people with the smallest distance (low-distant) to nuclear power, are inclined to support the state in matters of its development, whereas people, living in a state that does not support nuclear power and thus being "distant" from it, are convinced in a high risk of its use [19]. The human psyche is not able to assess all the risks of a process or phenomenon that is physically distant from an individual.

CLT clearly demonstrates that the level of public awareness in various aspects of energy is a key factor in its development as a whole. Currently, the relationship between the level of development of nuclear energy, state policy in this matter and public opinion is not well understood, especially among young people aged 21 to 35 years. The low level of elaboration of the problem determined the main goal of the research – studying the attitude of young people to nuclear energy in countries with different energy policies in order to identify the relationship between such factors as attitudes towards nuclear energy, risks, the level of state support in this matter and the awareness of the technical aspects of its use.

2 CONDITION OF NUCLEAR ENERGY INDUSTRY IN RUSSIA AND ITALY

Nowadays Italy is a leader in the development of renewable energy sources [20]. However, in the period from 1960s to 1980s the Italian government actively encouraged the development of nuclear energy as well. Due to Chernobyl disaster at 1986 which had taken place at the territory of the modern Ukraine, Italian energy policy significantly changed. In Russia, the development of nuclear energy didn't stop, neither after the Chernobyl disaster nor after serious political reforms in 1990s.

2.1 Nuclear energy industry in Russia

The Russian Federation is one of the leaders in nuclear energy usage and its development. Moreover, Russian nuclear industry is highly interested in several projects which are being implemented abroad such as construction of nuclear power plants and their further operation. In the next few decades Russian companies are planning to construct and build up more than 20 nuclear reactors in India, Pakistan, Algeria etc. These projects are worth more than $133 billion overall according to latest calculations made in 2017. Currently there are 35 operated nuclear reactors in Russia (Table 1).

Table 1: Operating nuclear reactors in the Russian Federation [21].

Reactor	Type V=PWR	Net MWe	Commercial operation	Licensed to, or scheduled close
Balakovo 1	V-320	988	5/86	2043
Balakovo 2	V-320	988	1/88	2033
Balakovo 3	V-320	988	4/89	2049
Balakovo 4	V-320	988	12/93	2053
Beloyarsk 3	BN-600 FBR	560	11/81	2025
Beloyarsk 4	BN-800 FBR	789	10/16	2056
Bilibino 2-4	LWGR EGP-6	11 x 3	12/74-1/77	Dec 2021
Kalinin 1	V-338	988	6/85	2045
Kalinin 2	V-338	988	3/87	2047
Kalinin 3	V-320	988	11/2005	2065
Kalinin 4	V-320	988	9/2012	2072
Kola 1	V-230	432	12/73	2033
Kola 2	V-230	411	2/75	2029
Kola 3	V-213	440	12/82	2027
Kola 4	V-213	440	12/84	2039
Kursk 1	RBMK	971	10/77	2022
Kursk 2	RBMK	971	8/79	2024
Kursk 3	RBMK	971	3/84	2029
Kursk 4	RBMK	925	2/86	2031
Leningrad 2	RBMK	971	2/76	2021
Leningrad 3	RBMK	971	6/80	2025
Leningrad 4	RBMK	925	8/81	2026
Leningrad II-1	V-491	1085	10/2018	2078?
Novovoronezh 4	V-179	385	3/73	2032
Novovoronezh 5	V-187	950	2/81	2035 potential
Novovoronezh II-1*	V-392M	1114	2/2017	2077
Smolensk 1	RBMK	925	9/83	2028
Smolensk 2	RBMK	925	7/85	2030
Smolensk 3	RBMK	925	1/90	2050
Rostov 1	V-320	990	3/2001	2030
Rostov 2	V-320	990	10/2010	2040
Rostov 3	V-320	1011	9/2015	2045
Rostov 4	V-320	1011	9/2018	
Total: 35	28,025 MWe			

Despite the fact that the level of nuclear energy development is considerably exceeds average point among other countries, the volume of its consumption is much lower in comparison to the fossil fuel usage. Fig. 1 provides the structure of primary energy consumption in Russia in 2017.

According to the Fig. 1, the large share in the primary energy consumption in Russia is oil and natural gas: 22% and 52% respectively, whereas the nuclear power share is only 7% – lower than coal consumption, which is 13%.Most of all, nuclear power plants

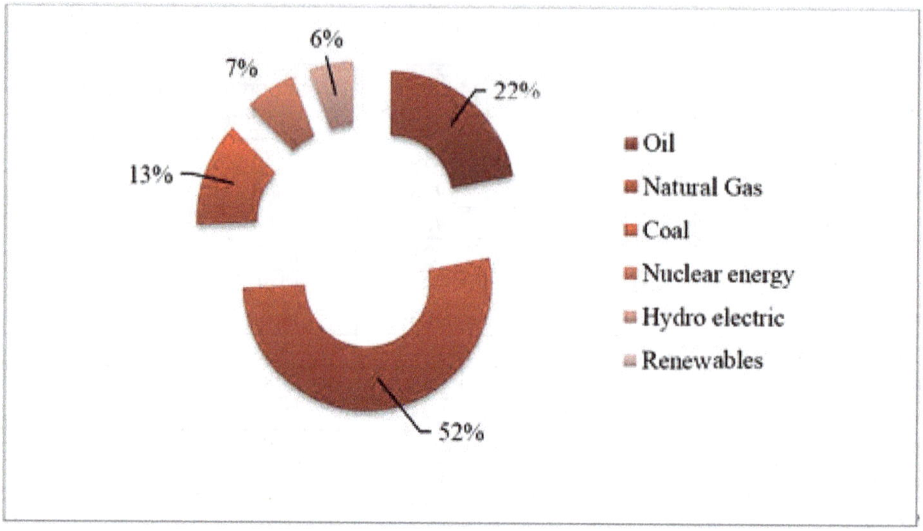

Figure 1: Primary energy consumption by fuel in Russia, 2017 (based on data from World Nuclear Association [21]).

are mainly located in the Western part of Russia due to the significant difference in the level of development among regions, their financing and population density.

Russia is also a leader in the development and producing of fast neutron reactors. This type of reactors has favourable environmental performance as that type of reactors can reduce the total radiotoxicity of nuclear waste [22]. Moreover, it has a standpoint of involving uranium 238 in the fuel chain, which is a strategic solution to the problem of providing nuclear fuel.

2.2 Nuclear energy industry in Italy

Nuclear energy in Italy was developing since 1958 till 1990s. In this period, 4 nuclear power plants were constructed and operated. Their total capacity was around 1423 MW (Table 2).

Table 2: Nuclear power plants operated in Italy since 1958 till the 1990s [23].

Reactor	Model	Net MWe	First power	Shutdown
Latina	GCR	153[h]	05/1963	12/1987
Garigliano	BWR	150	01/1964	03/1982
Enrico Fermi (Trino Vercellese)	PWR	260	10/1964	07/1990
Caorso	BWR	860	05/1978	07/1990
Montalto di Castro (Alto Lazio) 1&2	BWR	982 each	Cancelled	-
Total operated (4)	1423 MWe			

The Chernobyl disaster was the main reason for nuclear power plants' shutdown. In 1987, Italian government held a national referendum on the destiny of nuclear energy's future. Most citizens voted against of its development. Currently, Italy doesn't produce nuclear energy, however it is possible to import nuclear energy from other producers and take part in their construction within foreign nuclear projects.

The structure of primary energy consumption in Italy is quite similar to the Russian one. Consumption of imported fossil fuels takes the first and second place: natural gas consumption is about 40%, whereas oil consumption is 39%. Renewable energy is in the third place in terms of total energy consumption (Fig. 2). Both countries mainly use natural gas as a primary energy source: 40% in Italy and 52% in Russia.

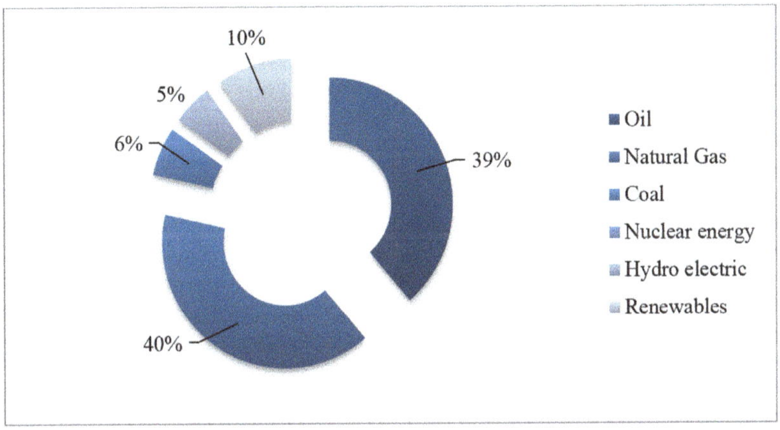

Figure 2: Primary energy consumption by fuel in Italy, 2017 (based on data from World Nuclear Association [23]).

It should be noted that the coal usage in Italy is significantly lower in comparison to Russia. Renewable energy in Russian primary energy consumption structure is less than 1% that provides low level of its development, while nuclear energy consumption is more than 7% (Fig. 1).

There is a completely different situation in Italy: while the generation of nuclear energy is absent, 10% of the total number of energy consumption is renewable energy sources, apart from hydroelectric.

3 RESEARCH METHODS

This research entails an international project studying students' attitude towards nuclear energy development. Authors carried out a survey in order to estimate the public acceptability of nuclear energy among young people of both counties. This paper analyses the questionnaire responses and provides crucial trends discovered within project study.

The survey was designed with aid of students of Ural Federal University: a technical section was made by the students of Power Engineering Institute and other section was provided by students of Graduate School of Economics and Management. Voluntary respondents were students of technical major studying at Ural Federal University named after the first president of Russia, B.N. Yeltsin (Ekaterinburg, Russia) and at University of Trento (Trento, Italy). The survey includes 8 multiply choice questions about different aspects of nuclear energy (Table 3).

Table 3: The structure of survey.

The question	Answer option
1. In your opinion, does the construction of nuclear power plants pose a potential risk to the environment of the region you currently live in?	1) Yes, it does 2) Possibly yes 3) Possibly not 4) No, it doesn't 5) It is hard to answer
2. In your opinion, could the nuclear energy be the part of the "green" energy?	1) Yes, it does 2) Possibly yes 3) Possibly not 4) No, it doesn't 5) It is hard to answer
3. What is the main benefit of producing the nuclear energy?	1) Ecological safety 2) Low expenses on producing 3) Low expenses on distributing 4) Effectiveness of energy supply 5) Low level of usage of fossil fuels 5) Long-term usage of nuclear power plants 6) Other (please, write you option)
4. Do you support the government's attitude towards the nuclear energy in the country you currently live in?	1) Yes, I do 2) No, I don't 3) It is hard to answer
5. What is the purpose of using the cooling tower at nuclear power plants?	1) For releasing of toxic emissions 2) For waste recycling 3) For cooling water 4) For heating nuclear fuel 5) It is hard to answer
6. What type of fuels is used at nuclear power plants?	1) Uranium 2) Radium 3) Tritium 4) Thorium 5) Any radioactive substance
7. How to protect yourself from alpha-radiation?	1) To put on outwear 2) To run away from radiation source 3) To hide behind the concrete wall 4) To start using specific medication 5) It is hard to answer
8. In your opinion, is a person exposed to radiation dangerous?	1) Yes, he does 2) No, he doesn't 3) Depends on time of exposure 4) Depends on type of radiation 5) It is hard to answer

The survey includes several blocks that aids to not only reveal the attitude of students to nuclear energy, but also to check their level of awareness of the basic aspects of the use of this type of energy.

The first section (questions 1–4) was a collection of information about the general attitude of young people to nuclear energy.

The second section of the survey (questions 5–8) related to the basic theoretical aspects of nuclear energy (mainly, security and technological components).

The survey includes three main stages:

1. Carrying out the survey among Russian students within 15th December and 25th December 2018. Overall there were 62 respondents: 23 of them are students of Civil Engineering Department, 39 of them are students of Power Engineering Institute. Questionnaire method: online survey.
2. Carrying out the survey among Italian Students in the period from 21st December to 23rd December 2018. There were 50 respondents from Department of Civil, Mechanical and Environmental Engineering. Questionnaire method: traditional survey.
3. Analysis of collected data using MS Office software and SPSS software.

The suggested survey is valid as a first stage of the general assessment of students' attitude towards nuclear energy in different countries and further may be more detailed developed and include additional questions.

4 RESULTS AND DISCUSSION

The study was based on the analysed data in order to confirm or refute the thesis about the correlation between the development of nuclear energy and public attitudes towards that in different countries (Italy and Russia). Both countries have completely opposite policies for the energy sector development: in Russia, nuclear energy shows an upward trend, while the Italian government has completely abandoned the construction and operation of nuclear power plants due to potential risks.

According to results of the first section, it might be concluded that the young people of both countries have more positive opinion about nuclear energy: more than half of all students surveyed do not consider nuclear energy dangerous (Table 4).

Table 4: The percentage of answers to the question "In your opinion, does the construction of nuclear power plants pose a potential risk to the environment of the region you currently live in?" (%).

Answer options	Ural Federal University		University of Trento
	Civil Engineering Department	Power Engineering Institute	
Yes, it does	30.43%	2.56%	22.00%
Possibly yes	17.39%	7.69%	4.00%
Possibly no	34.78%	56.41%	38.00%
No, it doesn't	0.00%	30.77%	26.00%
It is hard to answer	17.39%	2.56%	10.00%

The most negative evaluation of nuclear energy was given by students of the Civil Engineering Department at Ural Federal University that might be caused by lack of knowledge in this area or the specifics of the study major. At the same time, despite the negative attitude of the Italian government, most students do not consider nuclear energy to be potentially perilous. Many students mentioned the point, that in case of minimizing the impact of the human factor during nuclear plants operation, nuclear power might be the main perspective of the world energy sector in near future.

Despite the positive attitude of young people, most of Trento University students believe that nuclear energy cannot be attributed to the green energy. Fig. 3 shows a radar chart with a ranking of the answers to the question "In your opinion, could the nuclear energy be the part of the 'green' energy?"

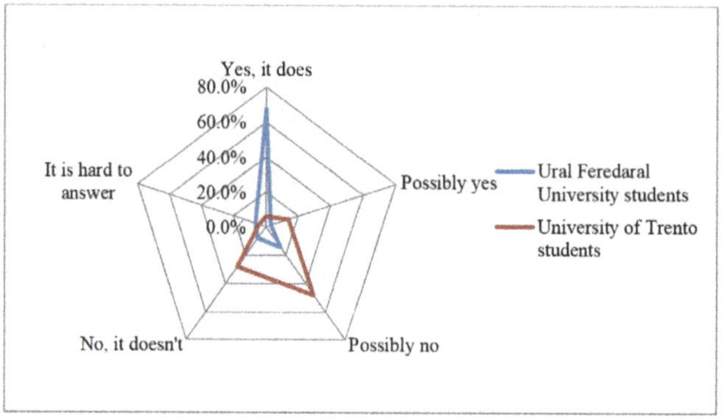

Figure 3: Petal chart of respondents' answers to the question about the possibility classifying nuclear energy as "green" one.

A possible reason for the significant difference in respondents' answer might be caused by fact that both the countries might have an absolutely different perception of nuclear energy.

Since the 1990s, the Italian government and the governments of other European countries (e.g. Austria, Czech Republic) have introduced a strict policy towards nuclear energy: in addition to the rejection of the construction and operation of nuclear power plants, the policy also has influenced the public attitude towards it and changed the public acceptability.

Students of the Russian University, on the contrary, believe that nuclear energy can be counted as a green one, and be part of environmentally safe energy sector, which is also due to the dissemination of information about the positive factors of its use (for instance, long-term perspective of its production and safety for the environment and people).

After the question about "green" energy, respondents had to answer the question related to the benefits of using nuclear power plants. According to the results, Russian students believe that the main benefit of nuclear energy is the environmental friendliness: more than 57% chose this answer, while the majority of Italian students mentioned the possibility of minimizing problems with the supply of electricity (82%). This ratio of responses clearly demonstrates the attitude of the population of both countries to the energy sector and its structure at all. It should be noted that none of the total number of respondents chose the option of saving fossil fuels. The final question of the first block was about the attitude of students to the current state policy on nuclear energy (Fig. 4). Respondents from both

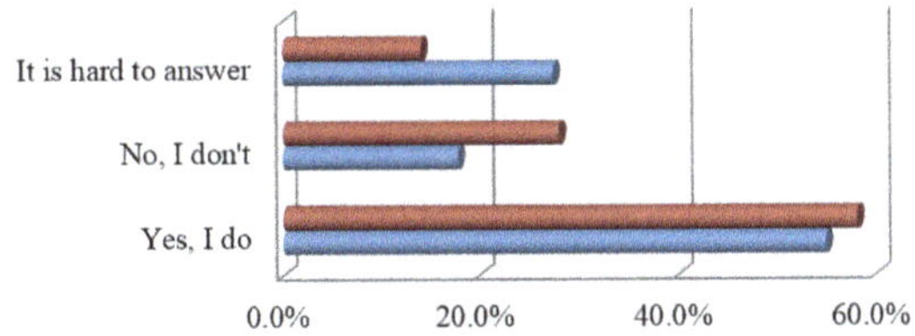

	Yes, I do	No, I don't	It is hard to answer
University of Trento students	58.00%	28.00%	14.00%
Ural Feredaral University students	54.8%	17.7%	27.4%

Figure 4: The ratio of students' answer about their approval of the governments' attitude to nuclear energy in the country they currently live in (%).

countries, approve of the government's policy according nuclear energy technologies, which is partially correlated with the answers to previous questions.

The second block of the survey (questions 5–8) related to the basic theoretical aspects of nuclear energy (mainly, security and technological components). Students from both countries demonstrated a high level of knowledge in the technical aspects of nuclear power plant's construction and operation. Fig. 5 shows the percentage of correct answers among students on questions 5–6 of the survey.

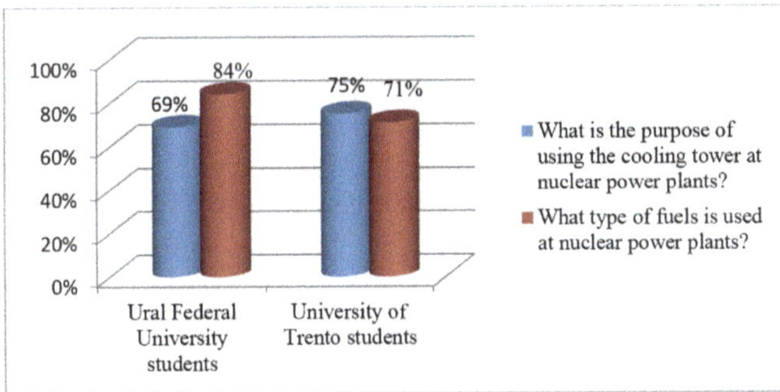

Figure 5: Correct students' answers to questions concerning technical aspects of nuclear plant's construction and operation (%).

However, energy students or civil engineering students usually get the basics of nuclear power plant operation in some courses during the first two years of study. A positive sign is that more than 50% of students from Russia and Italy gave the correct answer to the first two questions of the second block (the main and only purpose of using the cooling tower at nuclear power plants is water cooling process and the fuel used at nuclear power plants is uranium, respectively).

The reverse trend is observed in the issues of compliance with safety measures in case of emergencies. Most students of Ural Federal University gave the correct answer to the 7th question of the survey "How to protect yourself against alpha radiation?" (87.5%), whereas only 11% of Italian students are aware of the need to wear outerwear [24]. Moreover, 64% of Italian students preferred the answer "to run away from the radiation source". The main reason for this response ratio is that in Italy nuclear energy sector stopped its development more than 29 years ago, which causes the lack of need for training of this kind of skills. In Russia, this is a mandatory part of the subjects related to the safety of life.

The 8th question of the survey was "In your opinion, is a person exposed to radiation dangerous?". The main purpose of this question was to evaluate students' understanding of the effects of radiation on the human body. Table 5 shows the percentage of students in both countries who answered this question.

Table 5: The percentage of answers to the question "In your opinion, is a person exposed to radiation dangerous?".

Answer option	Ural Federal University	University of Trento (DICAM)
Yes, it does	27.42%	68.00%
No, it doesn't	51.61%	4.00%
Depends on time of exposure	9.68%	12.00%
Depends on type of radiation	1.61%	2.00%
It is hard to answer	12.90%	14.00%

Only 4% of Italian students gave the correct answer to the question, compared with 51.6% of Russian students: a person exposed to radiation is not a danger to others [25]. The possible reason for the results may coincide with the conclusion of the analysis of the answers to question 7. In general, a partial correlation was revealed between the use of nuclear power plants in a single state and the attitude of the population to this energy sector.

A significant part of Italian students, despite the relatively positive attitude to nuclear energy, supports the ban on the construction and operation of nuclear power plants in the Italian Republic, arguing for the need to develop renewable energy sources. Another identified feature is the high level of awareness of students of the University of Trento in the technological aspects of nuclear plant, but low knowledge in the field of security. Russian students living in the immediate vicinity of the Beloyarskaya Nuclear Power Plant demonstrated the relevant results: a positive attitude to nuclear energy, knowledge of the basic principles of its operation and safety.

According to a number of studies, the growth of loyalty to nuclear power in European countries was observed in 1990–2000 [26]. After this period, residents of most European countries have significantly changed their attitude, especially after the accident at the Fukushima nuclear power plant. In the course of the study conducted by the authors, it was expected that Italian students would demonstrate a low level of perception of nuclear energy, which would be due primarily to the attitude of the government [27]. Students of the University of Trento have no negative perceptions, as well as Russian students, but in general they do not support the initiative of construction and operation of nuclear power plants in territory of Italy. A possible reason may be that in Italy more attention is paid to renewable energy sources, which are actively developing in almost all parts of the country. The reason is public attitude: there are certain values for the maintenance of bioenergy among people, and nuclear energy is not perceived as a possible alternative. Moreover, another identified trend is the relationship between the age of respondents, their knowledge about nuclear power and their attitude to it. The authors found that young people, who, for the most part, possess a sufficient level of knowledge about nuclear power and are aware of the basic principles of nuclear power plants, are more loyal to the construction of additional nuclear power plants or power units to them. The revealed relationship between the level of knowledge and attitude leads to the conclusion that in the future the younger generation may be more loyal to nuclear energy, which will increase the pace of its development around the world.

5 CONCLUSIONS

In general, the results show that the younger generation of both countries is quite loyal to nuclear power: many understand its prospects in the energy sector and objectively assess the possibility of risks. Despite this, Italian students do not support the idea of building and renewing the use of nuclear energy in Italy, which is an element of public opposition to the whole idea of the development of nuclear energy. Meanwhile, public approval in such matters plays a key role, since it contributes to the formation of the necessary understanding of the use of certain technologies.

ACKNOWLEDGEMENT

This research was supported by Act 211 Government of the Russian Federation, contract № 02.A03.21.0006.

REFERENCES

[1] Rada, E.C., Special waste valorization and renewable energy generation under a circular economy: Which priorities? *WIT Transactions on Ecology and the Environment*, vol. 222, WIT Press: Southampton and Boston, pp. 145–157, 2018.

[2] Gitelman, L., Magaril, E., Kozhevnikov, M. & Rada E.C., Rational behavior of an enterprise in the energy market in a circular economy. *Resources*, **11**(7), p. 2165, 2019.

[3] European Environment Agency. Nuclear Waste Production, 2018. http://themes.eea.europa.eu/Sectorsandactivities/energy/indicators/EN13. Accessed on: 14 Apr. 2019.

[4] Hubbel, M.W., *The Fundamentals of Nuclear Power Generation: Questions & Answers*, AuthorHouse: Bloomington, 2011.

[5] Kyne, D., *Nuclear Power Plant Emergencies in the USA: Managing Risks, Demographics and Response*, Springer International Publishing AG: Switzerland, 2017.

[6] Georgios, K.P &, Dimitrios, N. (eds), *Nuclear Power Plants*, Nova Science Publishers: New York, 2011.

[7] Budnitz, R.J., Rogner, H.-H. & Shihab-Eldin, A., Expansion of nuclear power technology to new countries: SMRs, safety culture issues, and the need for an improved international safety regime. *Energy Policy*, **119**, pp. 535–544, 2018.

[8] Renewable Resources Coalition. Nuclear Energy: Pros and Cons, 2018. www.renewableresourcescoalition.org/nuclear-energy-pros-cons/. Accessed on: 15 May 2019.

[9] Antaki, G. & Gilada, R., *Nuclear Power Plant Safety and Mechanical Integrity: Design and Operability of Mechanical Systems, Equipment and Supporting Structures*, Elsevier: New York, 2015.

[10] Alam, F., Sarkar, R. & Chowdhury, H., Nuclear power plants in emerging economies and human resource development: A review. *Energy Procedia*, **160**, pp. 3–10, 2019.

[11] Verbruggen, A., Laes, E. & Lemmens, S., Assessment of the actual sustainability of nuclear fission power. *Renewable and Sustainable Energy Reviews*, **32**, pp. 16–28, 2014.

[12] Hugo, V.J. & Gertman, D.I., A method to select human-system interfaces for nuclear power plants. *Nuclear Engineering and Technology*, **48**, pp. 89–97, 2016.
[13] Siegrist, M., Keller, C. & Cousin, M.E., Implicit attitudes toward nuclear power and mobile phone base stations: Support for the affect heuristic. *Risk Analysis*, **26**, pp. 1021–1029, 2006.
[14] Arikawa, H., Cao, Y. & Matsumoto, S., Attitudes toward nuclear power and energy-saving behaviour among Japanese households. *Energy Research and Social Science*, **2**, pp. 12–20, 2014.
[15] World Nuclear Association, Nuclear Power in The USA (updated in April 2019). www.world-nuclear.org/information-library/country-profiles/countries-t-z/usa-nuclear-power.aspx. Accessed on: 31 Mar. 2019.
[16] Assefa, G. & Frostell, B., Social sustainability and social acceptance in technology assessment: a case study of energy technologies. *Technology in Society*, **29**(1), pp. 63–78, 2007.
[17] Geng, L., Liu, T., Zhou, K. & Yang, G., Can power affect environmental risk attitude toward nuclear energy? *Energy Policy*, **113**, pp. 87–93, 2018.
[18] Lammers, J., Galinsky, A.D., Gordijn, E.H. & Otten, S., Power increases social distance. *Social Psychological and Personality Science*, **3**(3), pp. 282–290, 2012.
[19] Magee, J.C. & Galinsky, A.D., 8 social hierarchy: The self-reinforcing nature of power and status. *The Academy of Management Annals*, **2**(1), pp. 351–398, 2008.
[20] European Environment Agency, Italy GHG and Energy 2017 Country Profile, 2017. www.eea.europa.eu/themes/climate/trends-and-projections-in-europe/trends-and-projections-in-europe-2017/country-profiles-greenhouse-gases-and-energy/italy-ghg-and-energy-country-profile.pdf/view.
[21] World Nuclear Association, Nuclear Power in Russia, 2018. www.world-nuclear.org/information-library/country-profiles/countries-o-s/russia-nuclear-power.aspx. Accessed on: 5 Apr. 2019.
[22] Quora, Is BN-800 the Best Nuclear Reactor For Now? 2007. www.quora.com/Is-BN-800-the-best-nuclear-reactor-for-now. Accessed on: 15 May 2019.
[23] World Nuclear Association, Nuclear Power in Italy, 2018. www.world-nuclear.org/information-library/country-profiles/countries-g-n/italy.aspx. Accessed on: 6 Apr. 2019.
[24] Radiation Emergency Medical Management, U.S. Department of Health and Human Services, Personal Protective Equipment (PPE) in a Radiation Emergency. www.remm.nlm.gov/radiation_ppe.htm. Accessed on: 9 Apr. 2019.
[25] Tsilikis, I., Pantos, I., Zouliati, I., Koutra, A., Kalinterakis, G. & Syllaios, A., Radiological risks from potential exposure of the population to radiation from orphan radioactive sources. *Health Physics*, **116**(5), pp. 715–719, 2019.
[26] Stoutenborough, W.J., Shelbi, G., Sturgess, G.S. & Vedlitz, A., Knowledge, risk and policy support: Public perception of nuclear power. *Energy Policy*, **62**, pp. 176–184, 2013.
[27] Huang, L., Zhou, Y., Han, Y., Hammitt, J.K., Bi, J. & Liu, Y., Effect of the Fukushima nuclear accident on the risk perception of residents near a nuclear power plant in China. *Proceedings of the National Academy of Sciences of the USA*, **110**(49), pp. 1972–1974, 2013.

PUBLIC PARTICIPATION IN ENVIRONMENTAL IMPACT ASSESSMENT AND ITS SUBSTANTIVE CONTRIBUTION TO ENVIRONMENTAL RISK MANAGEMENT: INSIGHTS FROM EIA PRACTITIONERS AND OTHER STAKEHOLDERS IN KENYA'S RENEWABLE ENERGY SUB-SECTOR

PHILIP M. OMENGE[1], GEORGE W. ESHIAMWATA[2],
STANLEY M. MAKINDI[3] & GILBERT O. OBWOYERE[1]
[1]Egerton University, Kenya
[2]The Kenya National Commission for UNESCO, Kenya
[3]Machakos University, Kenya

ABSTRACT

Comprehensive and transparent public participation during Environmental Impact Assessment (EIA) processes for renewable energy projects is vital in identifying, addressing and mitigating potential environmental risks associated with such renewable energy projects. Public participation during EIA is a vital platform where all stakeholders of a given renewable energy project contribute to addressing the environmental concerns of renewable energy projects, thereby contributing positively to informed environmental decisions that mitigate negative environmental impacts. This paper presents a comparative analysis of public participation practice during an EIA process in Kenya's renewable energy sub-sector vis-à-vis the international best practice operating principles. Further, the paper presents insights on the substantive contribution of public participation in environmental risk management based on questionnaire survey responses from EIA Practitioners and other stakeholders in Kenya's renewable energy sub-sector. Results show that public participation practice during EIA in Kenya loosely adheres to the international best practice operating principles. An analysis of stakeholder responses shows that public participation during EIA processes in the renewable energy sub-sector seldom supports decisions that result in environmental protection. Factors that contribute to Kenya's poor adherence to international best practice operating principles are discussed. Suggestions and recommendations on how to achieve a substantive contribution of public participation during EIA in Kenya's renewable energy projects in order to contribute to environmental risk management are presented.

Keywords: public participation, Environmental Impact Assessment, renewable energy, Kenya.

1 INTRODUCTION

Environmental Impact Assessment (EIA) is one of the processes that culminate to an environmental decision [1]–[3]. Most of EIA systems worldwide have embedded within them the requirement of public participation [4]–[11]. Whilst many definitions of public participation abound [4], [12], a scholarly discourse on the subject accentuate the rudiments of public participation as; consultative process, public involvement, inclusiveness, information sharing, transparency and influencing outcome of decisions [12]–[14]. Public participation in EIA process according to André et al. [15] "is the involvement of individuals and groups that are positively, or negatively affected by, or that are interested in, a proposed project, program, plan or policy, that is subject to a decision-making process". Public participation requirement during EIA process is underpinned in an array of international legal instruments such as the Aarhus Convention, United Nation Conference on Environment and Development, Convention on Environmental Impact Assessment in a Trans-boundary

WIT Transactions on Ecology and the Environment, Vol 237, © 2019 WIT Press
www.witpress.com, ISSN 1743-3541 (on-line)
doi:10.2495/ESUS190121

Context, North American Agreement on Environmental Cooperation, Principle 17 of the 1992 Rio Declaration on Environment and Development and Agenda 21 [1], [2], [9], [16]. In Kenya, public participation during EIA process is a constitutional requirement [17], [18]. The Environmental Management and Coordination Act, 1999 (Amended) 2015 which gives effects to article 69 of the Kenya Constitution 2010, makes public participation during EIA mandatory [3], [17], [19]–[21]. Public participation during EIA process is vital because it informs decision making, reduces conflicts, enhances transparency and accountability, builds trust, capture local and traditional knowledge, provide adequate opportunities to stakeholders to raise their concerns, educate stakeholders, increase awareness, build trust and legitimises public decisions [22]–[24]. Whereas public participation in EIA process is viewed as a means of nurturing a new ethos of environmental responsibility [25] for sustainable development [26], scholarly discourse continues on how it should be conducted [14]. Notwithstanding its importance [22], [23] its design and implementation remain contentious [14], [23], [24]. The objectives of this study were thus twofold, first to compare public participation practice during EIA process in Kenya vis-à-vis the international best practice operating principles. Secondly to find out the perceptions of EIA practitioners and other EIA stakeholders on the substantive contribution of public participation (during EIA process) to environmental decisions and environmental risk management in Kenya's renewable energy sub-sector.

2 METHODOLOGY

In line with the objectives, the research methodology was twofold. First, a comparative analysis of public participation practices during EIA process for renewable energy projects in Kenya vis-à-vis the international best practice operating principles of public participation. Secondly, a questionnaire survey among EIA Practitioners and other EIA stakeholders in Kenya's renewable energy sector. The international best practice operating principles of public participation adopted are according to André, et al. [15] while the Kenya public participation practice was as in Mwenda et al. [8], Mwenda and Kibutu [17], Kibutu and Mwenda [20]. Questionnaire respondents were sampled from licenced EIA practitioners in Kenya in the years 2018 while that for other stakeholders was sampled from the National Environment Management Authority (NEMA) Environmental Officers, the Ministry of Energy and civil society. The sampling method used for EIA practitioners was simple random sampling while that of other EIA stakeholders was purposeful sampling [27]. Example of questions in the questionnaire include, how effective is public participation during EIA in environmental decision making in renewable energy sub-sector? Does involvement of stakeholders during EIA contribute to informed environmental decision making in renewable energy sub-sector? Are public consultation sessions during EIA effective in contributing to informed environmental decision making in the renewable energy sector? How effective is public participation during EIAs in the renewable energy sub-sector in influencing sound decisions that contribute to environmental protection? In your EIA practice how effective do you think is the EIA tool in substantively contributing to informed environmental decisions as a result of implementation of a proposed renewable energy project? How effective are the stages of EIA process including that of public participation in contributing to informed environmental decisions? The responses were on a five-point Likert Scale (1–5) as follows: 1) very ineffective, 2) ineffective, 3) slightly effective, 4) effective, and 5) very effective. 200 responses from EIA practitioners, 14 from Ministry of Energy, 15 from Civil Society and 13 from NEMA were analysed.

3 RESULTS

3.1 Comparison of public participation practice during EIA process in Kenya with the international best practice operating principles

Table 1: Comparison of EIA International best practice operating principles of public participation with the EIA public participation practice in Kenya.

International best practice operating principles of public participation during EIA	Public participation practice during EIA process in Kenya's renewable energy projects
1) Initiated early and sustained: i) Public to be involved before major decisions are made ii) Public to be involved regularly in the EIA process	i) Public is involved at scoping, report preparation and report review stages ii) Public is involved by making contributing in three public meetings, during public hearing and by sending written and oral submissions on the EIA Study Report
2) Well planned and focused on negotiable issues: i) All impact assessment stakeholders should know the aims, rules, organization, procedure and expected outcomes of the public participation process undertaken ii) Emphasise understanding and respect for the values and interests of participants iii) Focus on negotiable issues relevant to decision making	i) Focus on methods of engaging the affected stakeholders ii) Focuses on explaining the project and its effects
3) Supportive to participants: i) Adequate diffusion of information on the proposal and on the public participation process ii) Equitable access to funding or financial assistance iii) Capacity-building, facilitation and assistance to groups who don't have the capacity to participate	i) Information on a proposed renewable energy project is only available at the website of the environmental agency and at the national and County office of the environmental agency where the proposed project is to be located ii) Diffusion of information on public participation process is limited to what is provided in EIA Regulations iii) There is no provision for funding support to enable economically disadvantaged stakeholders satisfactorily participate in the EIA process
4) Tiered and optimised: i) Public participation should occur at the most appropriate level of decision-making ii) The public should be invited to participate regularly, with emphasis on appropriate time for involvement iii) Optimization in time and space to ensure more willing participation	i) Public participate in public meetings before EIA report is compiled, in public hearing and send comments once the EIA report has been compiled ii) Public invited by notices, posters and radio announcement

Table 1: Continued.

International best practice operating principles of public participation during EIA	Public participation practice during EIA process in Kenya's renewable energy projects
5) Open and transparent: i) Access to all relevant information by all stakeholders ii) Provision of information and facilitation to ensure participation	i) Information on public participation is available at NEMA, website, print and electronic media ii) Information is in English language only
6) Context oriented: i) Be adapted to the social organization of the impacted communities, including the cultural, social, economic and political dimensions	The social organization of the impacted people is mostly ignored
7) Credible and rigorous: i) Adhere to established ethics, professional behaviour and moral obligations ii) Facilitation by a neutral facilitator	Facilitation during public meeting is by a Lead Expert while during public hearing is by NEMA official

3.2 Perceptions of EIA Practitioners and other EIA stakeholders on the effectiveness of public participation' substantive contribution to environmental decisions and environmental risk management in Kenya's renewable energy sub-sector

Table 2: Statistics of the respondents on the five-point Likert scale.

Category of respondents	Five-point Likert Scale					Total responses (n)
	Very effective (1)	Effective (2)	Slightly effective (3)	Ineffective (4)	Very ineffective (5)	
EIA Practitioner	12	29	29	95	35	200
Ministry of Energy	1	2	6	4	1	14
Civil Society	0	1	6	7	1	15
NEMA	1	2	6	4	0	13

4 DISCUSSION

4.1 Public participation practice during EIA in Kenya and its adherence to international best practice operating principles

We sort to determine how public participation practice during EIA process in Kenya's renewable energy sub-sector compares to the international best practice operating principles. Public participation during EIA process in Kenya is in three main stages namely at the scoping, storage, EIA report preparation stage and EIA report review stage. The first two stages are mainly sharing of proposed project information with the public while the third stage is mainly consultative in nature. Documented research has shown that information sharing form of public participation is passive public participation, it is viewed as non-participation, manipulative depicted by therapy as it is subsequent to decisions that have already been taken without inputs from the stakeholders [28]–[30]. Research has shown that public hearing is an ineffective form of public participation "it does not result in genuine participation, does not satisfy the public, seldom improve decisions as it does not include broad spectrum of the public but instead contributes to antagonising the public" [31]. The comparative analysis identified the following shortcomings of Kenya's public participation during EIA; public participation is not initiated early nor sustained throughout the EIA process because stakeholders such as community members of the proposed project site are not involved in early project stages of design, nor in the determination of project location further, stakeholders do not directly participate in the project approval stage. The practice is not well planned and does not focus on negotiable issues because it is organisationally deficient of a clear outline of what its aim is, rules and procedure to be followed and the expected outcome. It does not identify issues that stakeholders will negotiate on in order to aid decision making as stakeholders are viewed as a recipient of project information as opposed to equals capable of influencing project decision. Information diffusion on public participation and capacity building are both too limiting and prohibiting by design, location and language because capacity building for better public participation during EIA process is not actualised. Language used in notices, posters and radio announcement is commonly English which locks out many stakeholders. Information access is prohibitive as one will require access to internet and the requisite technical capacity to retrieve required information from relevant databases. The practice is not context oriented as cultural, social, economic and political dimensions are mostly ignored nor is it credible and rigorous as facilitators are interested parties and hence not neutral. Similar findings have been document in previous studies including Okello et al. [21] who concluded that public participation in Kenya's EIA process "is poor, particularly during the scoping, report review and follow-up stages" [8], who states that public participation within EIA process in Kenya is relatively low [32], who states that public participation in Kenya's EIA process is inadequate [33], who states that project "developers do not usually favour public participation, because they do not see the positive side of this process as a result, they are likely to hide information, or not clearly state data that may be controversial". Public participation in environmental decision making is both shaped by and, in many cases, constrained by the ways in which environmental issues, problems, and solutions are defined or framed through the strategic communication practices of the participants [34]. Exhaustive, inclusive and satisfactory public participation integrates local knowledge [2], broadens potential solutions [14], [35], [36], improves process outcomes [37] and avoids costly and time-consuming conflicts [38] thus guarantying access to justices in matters environment [14]. In line with the principles of informative, proactive and early involvement, the public should be involved as soon as value judgement becomes salient [13] in order to consider psychological and sociological understandings of risk [39]. These two principles underscore the importance of early public participation in the discourse of underlying assumptions and agenda setting as opposed to narrow predefined problems [40]. Effective public participation should be broad capturing representation of all affected public [13] for inclusivity, equitability, openness and transparency [17]. In order for public participation process to be considered truly imputable, the output of the participation should have a genuine impact on policy [13] otherwise the participation could be perceived as merely being used to legitimise already made decisions [28] as opposed to contributing to influencing sound environmental decision making [37]. Interactive participation is viewed as the only public participation method that enables stakeholders take control over decisions thus gaining a stake in mainstreaming structures and resources [29]. Interactive participation utilises systematic structured learning process from a multidisciplinary approach that enables stakeholders take control over decisions include resource usage. It enables stakeholders' participation in joint analysis, development of action plans and formation of and or strengthening of local institutions.

4.2 Public participation during EIA process in Kenya and substantive contribution to environmental decisions and environmental risk management

The substantive contribution of public participation during EIA process in environmental decision making and hence environmental risk management focuses on a number of issues including stakeholder participation [41]. It is on this basis we explored the perceptions of stakeholders who participate in EIA public participation process in Kenya's renewable energy sub-sector to establish how they perceive the effectiveness of their participation in substantively contributing to environmental decisions and hence environmental risk management. Results show a small percentage of between 6 and 8 (Figs 1–4) of respondents from each category perceive public participation during EIA process as being very effective in substantively contributing to environmental decisions for environmental risk management in the renewable energy sub-sector. A majority of the respondents from EIA practitioners (48%) (Fig. 1) and civil society (46%) (Fig. 3) and a significant percentage from NEMA (31%) (Fig. 4) and Ministry of Energy (29%) (Fig. 2) perceive public participation during EIA process as ineffective in substantively contributing to environmental decisions for environmental risk management in the renewable energy sub-sector. Research has shown that substantive contribution of public participation during EIA process in environmental decision making is influenced by and depended on local information and knowledge, incorporating experimental and value based knowledge and testing the robustness of information from other sources [24], [36], [42]. It could thus be argued that the observed perception of ineffectiveness of public participation's substantive contribution to environmental decisions could be attributed to inability of exhaustively harnessing local knowledge from local community stakeholders during public participation and incorporating the knowledge in environmental decisions. Research has also shown that unjust EIA procedures characterised by inequitable opportunity and freedom of affected communities to participate in the EIA process contribute to poor public participation of the affected stakeholders [43]. The outcomes of such an EIA procedure cannot support informed environmental decision but instead contribute to harming the environment [43]. Public participation is part of EIA process, poor public participation or lack of it amounts to unjust EIA procedures. Such unjust EIA procedures negate the substantive rationale of EIA which is to inform decision-making in order to mitigate negative environmental impacts [44] which will in turn contribute to environmental protection.

Whereas progress has been made in constitutionally and legislatively underpinning public participation in Kenya's EIA process, there is scanty information on actual execution. Constitutional and legislative loopholes are evident in Kenya's legal framework on public participations during EIA process. The constitution of Kenya falls short of making public participation mandatory in managing, conserving and protecting the environment but instead require the state to encourage public participation in the management, protection and conservation of the environment as captured in article 69(1)(d) [45]. To 'encourage' means to "give support, courage or hope" [46], waters down what could otherwise have been a mandatory constitutional requirement. The Environmental Management and Coordination Act, (EMCA) 1999 (Amended) 2015 and the Environmental (Impact Assessment and Audit) Regulation, 2003 provides for public participation during EIA process in Kenya [17]. These legislations fall short of defining the threshold required for public participation during EIA process to be considered imputable, credible and acceptable. Section 59 of EMCA provides for advertising an already prepared EIA study report in a newspaper, Kenya Gazette and radio and in the authority's website [47] as a form of public participation, this is passive participation. The EIA Regulation, 2003 which should spell out the nitty-gritties of public participation during EIA process to ensure that public participation is rigorous, exhaustive and all inclusive, instead reduces the process to "three public meetings" at strategic locations of the proposed project site [48]. The regulations only attempt to define how the public will be informed of the location and timings of the public meetings but fails to state how the process should be conducted to ensure credibility.

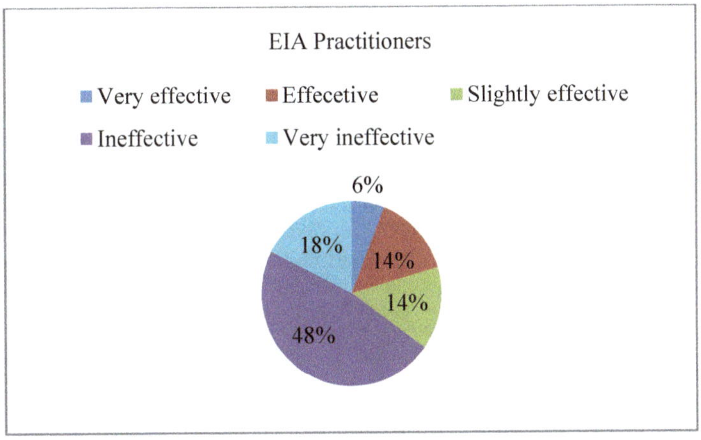

Figure 1: Perceptions of respondents of EIA practitioners on the effectiveness of public participation's substantive contribution to environmental decisions in Kenya's renewable energy sub-sector.

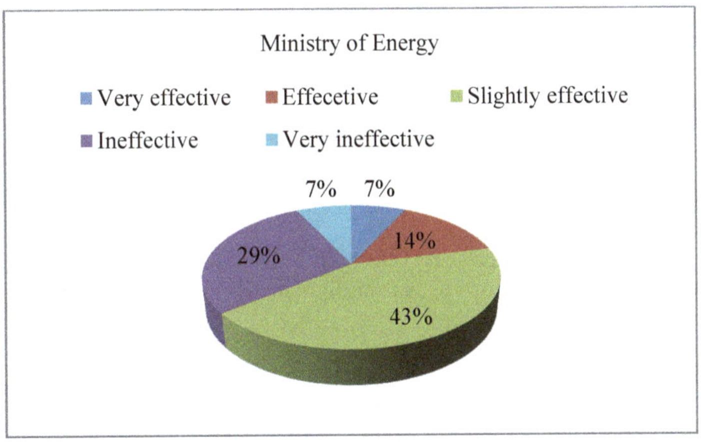

Figure 2: Perceptions of respondents from the ministry of Energy on the effectiveness of public participation's substantive contribution to environmental decisions in Kenya's renewable energy sub-sector.

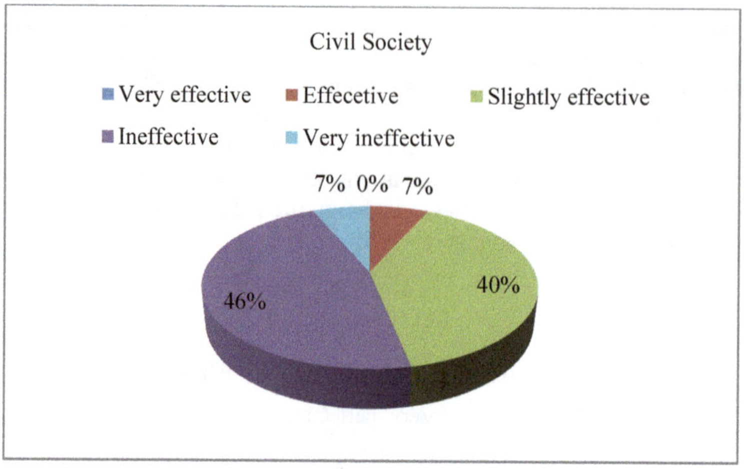

Figure 3: Perceptions of respondents from the civil society on the effectiveness of public participation's substantive contribution to environmental decisions in Kenya's renewable energy sub-sector.

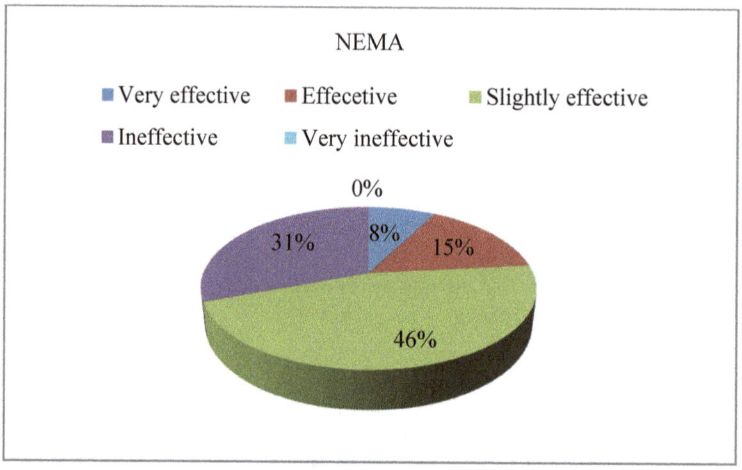

Figure 4: Perceptions of respondents from NEMA on the effectiveness of public participation's substantive contribution to environmental decisions in Kenya's renewable energy sub-sector.

5 CONCLUSION

Whilst the most appropriate method of public involvement depends on the specifics of any particular situation and that more knowledge-based decisions require lower levels of involvement than more value-based decisions, the forms of public participation used in the EIA process in Kenya are not satisfactorily interactive to a level where stakeholders take control over decisions thus gaining a stake in mainstreaming structures and resources. Public participation during EIA process in Kenya's renewable energy sub-sector therefore lacks the merits of the international best practice operating principles because it is not initiated early and is not sustained throughout the EIA process; it is not well planned and does not focus on negotiable issues; information diffusion and capacity building is too limiting and prohibiting by design, location and language; cultural, social, economic and political dimensions are mostly ignored and facilitators are interested parties most likely not neutral. The outcome of such a public participation for an EIA process seldom supports informed environmental decisions and hence cannot contribute to environmental risk management. The same finding is mirrored in the outcome of stakeholder perceptions (EIA practitioners, civil society, Ministry of Energy and NEMA respondents) where there is a general consensus that public participation during EIA process is ineffective in substantively contributing to informed decision making. This study has thus established that public participation during EIA process in Kenya's renewable energy sub-sector poorly adheres to the international best practice principles of public participation. The outcome of stakeholder perceptions on the substantive contribution of public participation in environmental decision making is that it rarely supports well informed decision-making that result in environmental protection in the renewable energy sub-sector in Kenya. The perception could be attributed to the poor adherence of the public participation practice to international best practice principles as seen in the shortcoming of the practice in Kenya. The outcomes thus show that for public participation to substantively inform decision making towards environmental protection, in order to minimise environmental risks, public participation during EIA in Kenya's renewable energy sub-sector should strictly adhere to the established international best practice operating principles.

6 RECOMMENDATIONS

To achieve substantive contribution of public participation during EIA in Kenya's renewable energy sub-sector, the current forms of public participation that are largely passive in nature should be substituted with interactive participation. The interactive stakeholder participation should begin from project conceptualization stage and sustained though the entire project cycle. To achieve this, it will be important for a review and broadening of the current EIA legislation in Kenya to provide for a standalone piece of legislation that is specific to public participation during environmental impact assessment process.

ACKNOWLEDGEMENTS

Special thanks to EIA practitioners, government agencies and civil society for responding to the questionnaire, Dr Jatin Nathwani, and the Global Change Initiative – Affordable Energy 4 Humanity (AE4H) which he spearheads at the Waterloo Institute for Sustainable Energy, University of Waterloo, Canada for the valuable advice and support.

REFERENCES

[1] Richardson, B.J. & Razzaque, J., Public participation in environmental decision-making. *Environmental Law for Sustainability*, **6**, pp. 165–194, 2006.

[2] Ocampo-Melgar, A., Sagaris, L., & Gironás, J., Experiences of voluntary early participation in environmental impact assessments in Chilean mining. *Environmental Impact Assessment Review*, **74**, pp. 43–53, 2019.

[3] Okello, N., Beevers, L., Douven, W. & Leentvaar, J., The doing and un-doing of public participation during environmental impact assessments in Kenya. *Impact Assessment and Project Appraisal*, **27**(3), pp. 217–226, 2009.

[4] Marzuki, A., A review on public participation in environmental impact assessment in Malaysia. *Theoretical and Empirical Researches in Urban Management*, **4**(12), pp. 126–36, 2009.

[5] Ruffeis, D., Loiskandl, W., Awulachew, S.B. & Boelee, E., Evaluation of the environmental policy and impact assessment process in Ethiopia. *Impact Assessment and Project Appraisal*, **28**(1), pp. 29–40, 2010.

[6] Marara, M., Okello, N., Kuhanwa, Z., Douven, W., Beevers, L. & Leentvaar, J., The importance of context in delivering effective EIA: Case studies from East Africa. *Environmental Impact Assessment Review*, **31**(3), pp. 286–296, 2011.

[7] Nadeem, O. & Fischer, T.B., An evaluation framework for effective public participation in EIA in Pakistan. *Environmental Impact Assessment Review*, **31**(1), pp. 36–47, 2011.

[8] Mwenda, A.N., Bregt, A.K., Ligtenberg, A. & Kibutu, T.N., Trends in consultation and public participation within environmental impact assessment in Kenya. *Impact Assessment and Project Appraisal,* **30**(2), pp. 130–135, 2012.

[9] Morgan, R.K., Environmental impact assessment: the state of the art. *Impact Assessment and Project Appraisal,* **30**(1), pp. 5–14, 2012.

[10] Rebelo, C. & Guerreiro, J., Comparative evaluation of the EIA systems in Kenya, Tanzania, Mozambique, South Africa, Angola, and the European Union. *Journal of Environmental Protection*, **8**(5), pp. 603–636, 2017.

[11] Glucker, A.N., Driessen, P.P., Kolhoff, A. & Runhaar, H.A., Public participation in environmental impact assessment: why, who and how? *Environmental Impact Assessment Review*, **1**(43), pp. 104–111, 2013.

[12] Slocum, R. & Thomas-Slayter, B., *Participation, Empowerment and Sustainable Development. Power, Process and Participation: Tools for Change.* Intermediate Technology Publications: London, pp. 3–8, 1995

[13] Rowe, G. & Frewer, L.J., Public participation methods: A framework for evaluation. *Science, Technology, & Human Values*, **25**(1), pp. 3–29, 2000.

[14] Hartley, N. & Wood, C., Public participation in environmental impact assessment: Implementing the Aarhus Convention. *Environmental Impact Assessment Review,* **25**(4), pp. 319–340, 2005.

[15] André, P., Enserink, B., Connor, D. & Croal, P., Public participation international best practice principles. *Special Publication Series,* **4**, 2006

[16] Kolhoff, A.J., Driessen, P.P. & Runhaar, H.A., Overcoming low EIA performance: A diagnostic tool for the deliberate development of EIA system capacities in low and middle income countries. *Environmental Impact Assessment Review*, **68**, pp. 98–108, 2018.

[17] Mwenda, A. & Kibutu, T.N., Implications of the new constitution on environmental management in Kenya. *Law, Environment & Development Journal*, **8**, p. 76, 2012.

[18] Kibugi, R., Development and the balancing of interests in Kenya. *The Balancing of Interests in Environmental Law in Africa*, p. 169, 2011.

[19] Kameri-Mbote, P., Strategic planning and implementation of public involvement in environmental decision-making as they relate to environmental impact assessment in Kenya. Nairobi, Kenya., *International Environmental Law Research Center Working Paper*, 2000-3, 2000.

[20] Kibutu, T.N. & Mwenda, A.N., Provision for environmental impact assessment (EIA) in Kenya's legislation: A review of the Environmental Management and Coordination Act (EMCA) and Environmental (Impact Assessment and Audit) Regulations (EIAAR). *Eastern Africa Journal of Humanities and Sciences*, **10**(2), pp. 1–3, 2010.

[21] Okello, N., Douven, W., Leentvaar, J. & Beevers, L., Breaking Kenyan barriers to public involvement in environmental impact assessment. *Proceedings of the Annual Conference of the International Association for Impact Assessment,* **19**, pp. 1–5, 2000.
[22] Innes, J.E. & Booher, D.E., Reframing public participation: Strategies for the 21st century. *Planning Theory & Practice*, **5**(4), pp. 419–436, 2004.
[23] Stewart, J.M. & Sinclair, A.J., Meaningful public participation in environmental assessment: Perspectives from Canadian participants, proponents, and government. *Journal of Environmental Assessment Policy and Management*, **9**(2), pp. 161–83, 2007.
[24] O'Faircheallaigh, C., Public participation and environmental impact assessment: Purposes, implications, and lessons for public policy making. *Environmental Impact Assessment Review*, **30**(1), pp. 19–27, 2010.
[25] Engel, J.R., The faith of democratic ecological citizenship. *The Hastings Center Report,* **28**(6), pp. 31–34, 1998.
[26] Chi, C.S., Xu, J. & Xue, L., Public participation in environmental impact assessment for public projects: A case of non-participation. *Journal of Environmental Planning and Management*, **57**(9), pp. 1422–1440, 2014.
[27] Kothari, C.R., *Research Methodology: Methods and Techniques*. New Age International, 2004.
[28] Cornwall, A., Unpacking 'participation': models, meanings and practices. *Community Development Journal*, **43**(3), pp. 269–283, 2008.
[29] Pretty, J.N., Participatory learning for sustainable agriculture. *World Development,* **23**(8), pp. 1247–1263, 1995.
[30] Arnstein, S.R., A ladder of citizen participation. *Journal of the American Institute of Planners,* **35**(4), pp. 216–224, 1969.
[31] Innes, J.E. & Booher, D.E., Reframing public participation: Strategies for the 21st century. *Planning Theory & Practice*, **5**(4), pp. 419–436, 2004.
[32] Kakonge, J., Environmental impact assessment: Why it fails in Kenya. *Pambazuka News*, 2015.
[33] Enríquez-de-Salamanca, Á., Stakeholders' manipulation of environmental impact assessment. *Environmental Impact Assessment Review*, **68**, pp. 10–18, 2018.
[34] Depoe, S.P., Delicath, J.W. & Elsenbeer, M.F. eds., *Communication and Public Participation in Environmental Decision Making*, Albany: SUNY Press, 2004.
[35] Vanclay, F., International principles for social impact assessment. *Impact Assessment and Project Appraisal*, **21**(1), pp. 5–12, 2003.
[36] Morrison-Saunders, A. & Early, G., What is necessary to ensure natural justice in environmental impact assessment decision-making? *Impact Assessment and Project Appraisal*, **26**(1), pp. 29–42, 2008.
[37] Sinclair, A.J., Diduck, A. & Fitzpatrick, P., Conceptualizing learning for sustainability through environmental assessment: Critical reflections on 15 years of research. *Environmental Impact Assessment Review*, **28**(7), pp. 415–428, 2008.
[38] Diduck, A. & Mitchell, B., Learning, public involvement and environmental assessment: A Canadian case study. *Journal of Environmental Assessment Policy and Management,* **5**(3), pp. 339–364, 2003.
[39] Renn, O., Webler, T., Rakel, H., Dienel, P. & Johnson, B., Public participation in decision making: A three-step procedure. *Policy Sciences*, **26**(3), pp. 189–214, 1993.
[40] Moffet, J., Environmental priority setting based on comparative risk and public input. *Canadian Public Administration*, **39**(3), pp. 362–385, 1996.

[41] Chanchitpricha, C., Bond, A. & Cashmore, M., Effectiveness criteria for measuring impact assessment tools. *SEA Implementation and Practice: Making an Impact*, **43**, pp. 1–13, 2011.

[42] Abaza, H., Bisset, R. & Sadler, B., Environmental impact assessment and strategic environmental assessment: towards an integrated approach. UNEP/Earthprint, 2004.

[43] Simpson, N.P. & Basta, C., Sufficiently capable for effective participation in environmental impact assessment? *Environmental Impact Assessment Review*, **70**, pp. 57–70, 2018.

[44] Loomis. J.J. & Dziedzic, M., Evaluating EIA systems' effectiveness: a state of the art. *Environmental Impact Assessment Review*, **68**, pp. 29–37, 2018.

[45] *Constitution of Kenya*, Government printer, Nairobi, 2010.

[46] Hornby, A.S., Cowie, A,P., Gimson, A.C. & Hornby, A.S., *Oxford Advanced Learner's Dictionary of Current English*, Oxford: Oxford University Press, 1974.

[47] The Environmental Management and Co-ordination (Amendment) Act, 2015, Kenya Gazette Supplement No. 74, Acts No. 5, 2015.

[48] Kenya, Environmental (Impact Assessment and Audit) Regulations, Legal Notice No. 101, Kenya Gazette Supplement No. 56, Legislative Supplement No. 31, 2003.

COMMITMENT TO USE SUSTAINABLE ENERGY IN A WORST-CASE SCENARIO: SAN QUINTÍN AGRICULTURAL REGION OF BAJA CALIFORNIA, MEXICO

ISAAC AZUZ-ADEATH & ROBERTO ARREOLA-SEVILLA
Center for Innovation and Design in Engineering (CEID), CETYS University, Mexico

ABSTRACT

The agricultural region of San Quintín in the Mexican state of Baja California is a 10,000 km^2 semi-arid coastal plain located 300 km south of the US–Mexico border. It is one of the most important agricultural areas of the country thanks to the use of high technology and to the quality of its products. The San Quintín region is the main national exporter of berries, and 20% of the tomato exported to the USA is produced there. Economic development and social welfare options have been factors for population attraction (approximately 30,000 inhabitants in 2010, plus 5,000 temporal workers every year), which, in addition to the natural conditions of the area (no surficial rivers, 150 mm/year of rain), has led to critical hydrological stress (63% of the aquifers are overexploited or salinized). Due to clean water scarcity and long-term drought, cultivated areas have been reduced from 25,000 ha in the 1980s to 8,000 in the present. Despite adverse conditions, San Quintín producers are committed to sustainability and are looking to be the first "green" agricultural district in the country by reducing their CO2 and hydrologic footprint during production. At least 67 desalination plants that process water from the aquifers operate in the region, and the first plant that will desalinize seawater for agricultural use is in process. Even with a government subsidy of 90% of the cost of energy used in the field, producers are considering the use of alternative energy for their desalination plants. This paper shows the strategies and cost of implementation of a renewable energy network for use in desalination plants in the agricultural region of San Quintín.

Keywords: sustainable energy, desalination plants, agricultural production, San Quintín, Mexico.

1 INTRODUCTION

The border region of California (USA) and Baja California (Mexico) is one of the most dynamic economic areas in the world. During the second quarter of 2018, Baja California experienced an economic growth of 3.6%, thus becoming the state with the highest growth rate of the northern border of Mexico. In the same period, the primary economic activities in which agriculture is included, registered an increase of 9.8% [1].

The San Quintín Valley is located 300 km south of the US–Mexico border and is part of the irrigation district of Ensenada. It is the largest agricultural valley in the state of Baja California, which is a state that functions with local water sources. With an approximate area of 10,000 km^2 and a semi-arid climate, this area sits on the coastal plain of the Pacific littoral.

The most important agricultural crops of the San Quintín Valley are: tomatoes, strawberries, berries, grapes, cucumber and chili. This valley produces 20% of the Mexican tomatoes that are exported to the US market and is the main Mexican exporter of berries. The agricultural activities of the region, in addition to contributing to the state economy, employ thousands of workers and have contributed decisively to improving the living conditions of its inhabitants.

Despite the marked decrease in cultivable areas, the economic value of production shows a continuous increasing trend, as shown in Fig. 1.

www.witpress.com, ISSN 1743-3541 (on-line)
doi:10.2495/ESUS190131

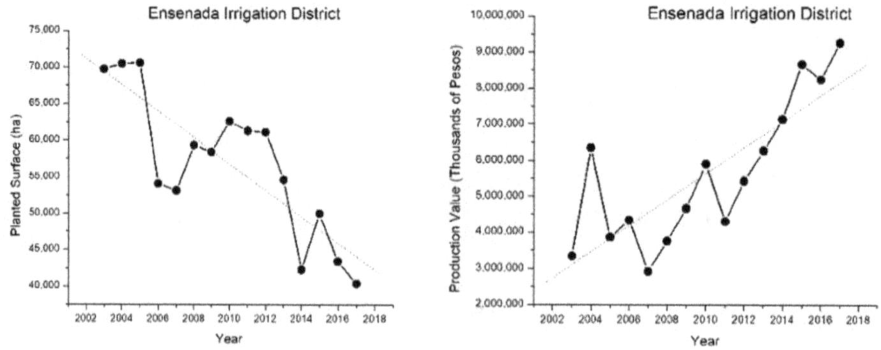

Figure 1: Historic planted surface (left) and production value (right) in the Ensenada irrigation district [2].

The cause of decreased cultivation areas is water scarcity in the region, which is due to lack of rainfall, surface rivers and a decrease of aquifer levels, as well as salinization of aquifers. Production value increases according to market conditions, quality of the products, intensive cultivation and use of agricultural technology (e.g. water efficiency strategies, extensive use of fertilizers and agrochemicals, better agricultural system management, etc.). Combined with lack of water, global warming has caused a considerable increase of atmospheric temperature [3]. Fig. 2 shows the behavior of some climatic variables (i.e. rain, maximum and minimum temperature) in the state of Baja California.

The San Quintín Valley has an annual average rainfall of 200 mm and there are no permanent rivers. Furthermore, out of the eight aquifers, three are overexploited, and three are salinized. In this scenario, the only option available to obtain water for agricultural purposes is the use of desalination plants. In 2017, at least 67 desalination plants are operating and registered at CONAGUA, the national water authority, and all of them use water from the salinized aquifers because they are cheaper to process than seawater due to their lower content of salt. It is difficult knowing the total water extraction capacity of the region because desalination plants are private, and are installed, operated and maintained by ranches and farms themselves. However, at least one of them has a capacity to process 24 l/s. For comparative purposes, the El Paso water plant, which is the world's largest inland desalination facility, can produce up to 122 million liters per day [5], while in the San Quintín Valley, there is only one plant that produces 2 million liters per day.

The exact amount of energy needed to operate the desalination plants in San Quintín valley is unknown, but demand for electric power must be very large. All facilities are connected to the regional power grid (operated by the federal government through Comisión Federal de Electricidad, or the Federal Electricity Commission), because the federal government subsidizes more than 90% of the cost of energy used for agricultural purposes. The Special Energy Program for Agricultural Use establishes that "if the regular cost of energy is 7.24 pesos per Kwh, and the farmer signs the agreement, the daytime cost will be 0.62 pesos per kWh and the nighttime cost will be 0.31 pesos per kWh".

The state of Baja California has great potential for harnessing renewable energies in Mexico. In San Quintín Valley, the solar and wind potential was deemed high by the Ministry of Energy. The potential of solar energy ranges from 5500 to 5749 Wh/m^2/day, and wind velocity ranges from 6 to 7.25 m/s, with plant or power factor of 20 to 30%, ideal for the

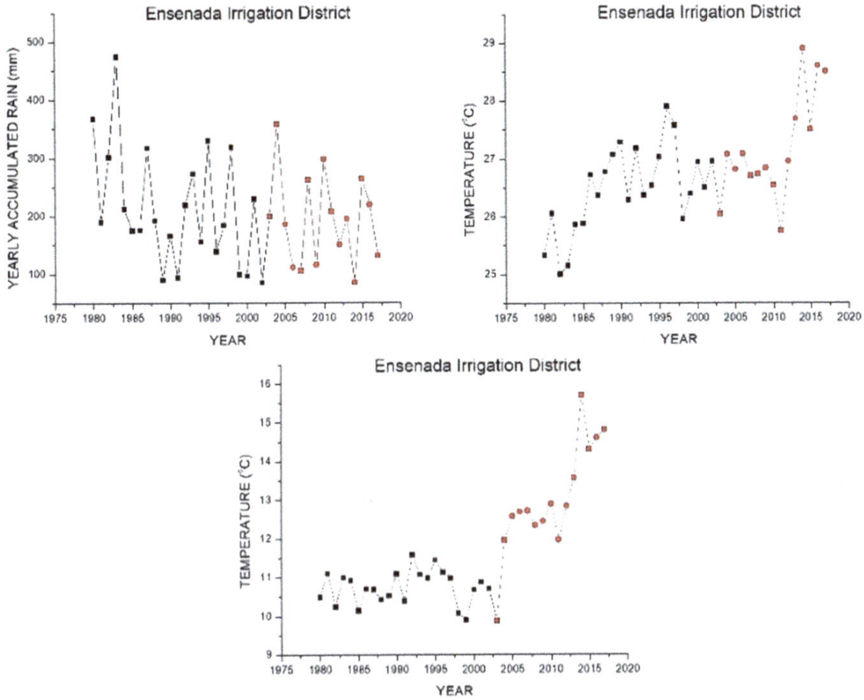

Figure 2: Historic accumulated rain (top left), maximum temperature (top right) and minimum temperature (bottom). The portion of the time series corresponding to the data period shown in Fig. 1 is marked in red [4].

installation of wind farms [6]. The coastline of this region (from Punta Colonet to El Rosario) is approximately 135 km long and has direct access to the Pacific Ocean. Even without having quantified wave power energy locally, it can be considered high, and it can be a potential source of electric power generation. Moreover, waste and debris from agricultural processes could be used as biofuel, since it is generated in large quantities during harvest.

It is a complex scenario: an agricultural region focused on exporting products with high added value, limited by the availability of water but that has implemented desalination at an industrial level, uses electric power from the grid thanks to large government subsidy, and is rich in natural conditions that give it unique qualities for exploitation of renewable energy sources. The only way to achieve sustainable development is if producers commit to develop a plan to gradually incorporate clean energy. Agricultural producers are aware of the problems that could arise from the introduction of tariff barriers or dumping policies to their product exports due to government subsidies and have decided to become the first "green" or sustainable agricultural district in Mexico. This paper shows the strategies and cost of implementation of a renewable energy network to be used in desalinization plants already installed in the agricultural region of San Quintin, Baja California, Mexico.

2 STUDY AREA AND METHODOLOGY

The San Quintín Valley is located in north western Mexico (see Fig. 3), in the state of Baja California, inside the Ensenada Municipality. Administratively, it comprises four delegations: Camalú, Vicente Guerrero, San Quintín and El Rosario, with a total surface of 10,311 km^2. In 2010, the total permanent population of the San Quintín Valley was 26,557 inhabitants (roughly 10% of the county's population), but the fluctuating population associated with agricultural harvesting processes can represent up to 50% of the permanent population [7]. Many of these temporary workers decide to stay and live in the region, thus increasing the need for basic services and infrastructure, particularly water and energy. The main agricultural products produced in the region are: tomatoes, strawberries, berries, grapes, cucumbers, and chilies; most of them are produced for exportation.

Figure 3: Study area, San Quintín region in Mexico. *(Maps Sources: IMIP, 2018; INEGI, 2018 and Addicted04, MEX orthographic, CC BY-SA 3.0. [8], [9].)*

In order to define the energy and hydric requirements of the San Quintín Valley, this study uses the two main crops in the area as a reference (tomatoes and strawberries). These crops represent approximately 63% of the planted surface in the valley and 64% of the total production value. Both products are cultivated under controlled conditions (i.e. high-tech greenhouses with water efficiency systems) in areas of 0.75 ha (1 greenhouse facility).

Without specific information about the type, operation conditions and capacity of all desalination plants in the region, this study is based on field information and personal communications with farmers, and uses the following assumptions: (i) all of the desalination plants use reverse osmosis, (ii) the smallest capacity plants produce 0.37 l/s, (iii) the largest capacity plants produce 24 l/s, (iv) based on information of three medium size desalination plants the energy efficiency was between 63% to 67% and, (v) the average monthly energy consumption for the desalination plant was obtained using one year of measured data (kWh) from two medium size plants. The cost of energy for the desalination process also uses three scenarios: regular cost, diurnal energy cost with subsidy and nocturnal energy cost with subsidy.

To define the specific water needs for each type of crop, this study follows the methodology proposed by Fuentes-Yagüe and García-Legaspi [10], which is widely used in Mexico. This methodology considers the duration of the different growth stages of each crop, the type of soil, the local monthly mean surface atmospheric temperature, and plant and crop evapotranspiration quantities. The local values for some of the aforementioned variables were obtained from the Camalú station [11], the governmental measurement station closest to the study area. The crop coefficients (Kc) used in this study are shown in Table 1.

Total water required was calculated considering the water needs and planted area of both crops. Using this information and the scenarios proposed in the previous paragraphs regarding types of desalination plants and energy cost, the current total energy needs for the desalination plants used for agriculture in the San Quintín Valley were stablished.

Table 1: Single (time-average) crop coefficients (Kc) used in this study [12].

Crop stages	Tomatoes		Strawberries	
	Duration (days)	Kc	Duration (days)	Kc
1	35	0.60	20	0.40
2	40	1.15	50	0.85
3	50	0.80	75	0.75
4	30	0.60	60	0.20
Total	155		205	

Different scenarios regarding the move towards use of renewable energy sources (solar photovoltaic and wind) were established considering the water and energy needs of the valley, including the time needed to achieve economic equilibrium.

3 RESULTS: HYDROLOGICAL AND ENERGY NEEDS OF THE REGION

The six main agricultural products of the San Quintín Valley represent less than 20% of the total planted surface in the irrigation district of Ensenada. However, in 2017 their production value accounts for 76% of the total income. Due to water availability restrictions, from 2003 to 2017, the planted surface in the San Quintín Valley experienced a decrease of 30%, going from roughly nine thousand hectares to seven thousand. Despite this, the value of production tripled, especially in the case of strawberry crops (see Table 2).

Table 2: San Quintín Valley main crops information 2003–2017 [13].

Crop	Year 2003			Year 2017		
	Planted Surface (ha)	Production (ton)	Production value (thousands of Pesos)	Planted surface (ha)	Production (ton)	Production value (thousands of Pesos)
Chili	684.0	14,269.4	62,406.7	815.0	22,516.4	247,626.0
Raspberry	25.0	205.0	5,798.4	829.0	12,048.5	1,469,405.8
Strawberry	764.0	33,941.1	324,979.2	1,869.0	91,660.3	2,050,893.3
Cucumber	1,523.5	48,546.2	283,073.4	712.6	54,979.1	718,293.2
Tomato	5,608.2	250,446.6	1,575,898.9	2,310.5	179,005.7	2,419,473.4
Grape	577.0	1,249.5	6,473.1	450.7	3,385.7	34,085.0
Blackberry	2.0	–	–	56.0	721.5	76,951.5
Total	9,183.7	348,657.8	2,258,629.8	7,042.8	364,317.2	7,016,728.2

Four stages of plant growth, as well as corresponding evapotranspiration (Evapo) were considered in order to calculate water needs for the two selected crops. This data provided the information necessary to establish the water needs of each cultivated hectare. Table 3 shows the results obtained from tomato crops.

Results show that one hectare of tomato crops requires, from sowing to harvest, 565.72 mm of total water, which represents 5,657 cubic meters or 5.6 million liters. Similar calculations were done for strawberry crops, resulting in a total of 3,352 m^3 per hectare.

Table 3: Example of water needs calculations for tomato crops.

Crop stage	Date	Time duration (days)	Crop evapo (mm)	Monthly water needs (mm)
1	1–30 April	30	2.74	82.08
	1–5 May	5	2.83	154.96
2	5–31 May	26	5.42	
	1–14 June	14	5.38	135.25
3	15–30 June	16	3.74	
	1–31 July	31	3.36	104.16
	1–3 August	3	3.50	84.09
4	4–31 August	28	2.63	
	1–2 September	2	2.59	5.18
Total		155		565.72

Considering the total area occupied by both crops (tomatoes and strawberries) in the San Quintín Valley, the total amount of water required would be 5,657 m^3 x 2,310.5 ha + 3,352.5 m^3 x 1,869 ha = 19.3 million cubic meters per agricultural cycle.

This amount of water must be obtained through desalination, which requires a large amount of energy. There is no single desalination plant that supplies the entire area; instead, there is a large number of plants in the area that are connected to the power grid, which in turn increase the operating costs because they form a dispersed distribution network. During this study we found 67 registered desalination plants with different capacities, all of them using the reverse osmosis method. Many farmers in the area consider the information related to water desalination and energy cost to be a confidential matter, however, the data related to an average or "type" plant was obtained for this study. This study uses the following data to make generalizations of the entire agricultural production area of the San Quintín Valley. Changes in desalination plant efficiency could modify these calculations:

- Water treated (l/h) = 4,051
- Water produced (l/h) = 1,326
- Water rejected (l/h) = 2,725
- Desalinization efficiency = 33%
- Initial water quality/salinity (ppm) = 18,000
- Final water quality/salinity (ppm) = 1.9
- Average monthly energy consumption (kWh) = 6,840

In the case of tomato crops (see Table 3), the total cultivation cycle lasts 155 days, which corresponds to approximately 5.2 months, and requires 13,070.5 million liters of water (total cultivation time times total cultivated area); for strawberry crop the cultivation cycle lasts 6.8 months, and require 6,262.8 million liters of total water. If only one desalination plant supplies the entire region, it should produce 2,500 million liters per month. Using the above information to calculate the energy consumption of a "type" desalination plant, roughly 1 million liters/month require 6.3 kW of energy, so the cultivated area corresponding to tomato crops in the San Quintin Valley would need (2,514 million l/month) x (5.2 months) x (6.3 kW/million l) = 82,359 kW or 82.4 MW of energy for irrigation. Similar calculations were done for strawberry crops need (921 million l/month x 6.8 months x 6.3 kW/million l), resulting in 39.5 MW of energy consumption needed for irrigation. Therefore, the total energy demand to supply desalinate water for the two crops analyzed in the study region during one agricultural cycle would be 121.9 MW.

In Mexico there are different elements involved in the calculations of electric energy cost:

1. Geography. The country is divided into 8 regions according to income of the population. Different prices exist for each region.
2. Seasonality. The year is divided into summer and winter seasons. Different prices exist for each season.
3. Type of use and amounts used. Four different uses are defined with several categories each, according to the amount of energy used in each one: (a) residential, (b) public services, (c) agricultural irrigation and (d) general purposes. Different prices exist for each type of use and amount of energy used.

Considering the "type" conditions, the amount of Mexican pesos paid for each kWh of energy in the agricultural sector (without a governmental subsidy) is \$7.24 Mexican pesos. Considering this, the total amount of money required to pay for the energy needed to desalinate water for one agricultural cycle of tomato and strawberry crops in San Quintin, would be 636 million Mexican pesos (approximately 35.3 million American dollars). With the government subsidy (Special Energy Program for Agricultural Use) the total cost goes down to between 27.2 and 54.4 million Mexican pesos (considering the different costs for diurnal and nocturnal energy) or 1.5 to 3.0 million American dollars.

The potential of wind and solar energy in San Quintin region is very high [6], the agricultural region is located near the coastline (between 1 and 10 km) directly in front of the open ocean (Pacific Ocean), and it has available enough land for the installation of large energy facilities (wind turbines, solar systems, etc.).

To meet the energy demand associated with desalination process for the crops analyzed in this study (approximately 122 MW per agricultural cycle), only two general scenarios were considered: exclusive use of photovoltaic solar energy and exclusive use of terrestrial and marine wind energy. In the first case, considering solar panels with generation capacity of 250 W (common capacity of existing solar panels in the Mexican market), 488,000 solar panels would be needed. In the second scenario, using operational information from wind turbines in Europe [14] (in average 5,600 kW per marine turbine), the number of wind turbines required to supply the energy demand in the agricultural region, goes from 22 (marine) to 44 (terrestrial).

4 CONCLUSIONS

This study confirms the premise that "water availability and growth are interconnected" [15]. For the San Quintin region, the only option available to exist as a successful agricultural area is desalination. The two crops analyzed in this study require nearly 20 million cubic meters of water per agricultural cycle, which are obtained through desalination. The calculations show that regardless of the commitment made by farmers to obtain energy from renewable sources, the cost of this transition will be very high.

The water-energy-food nexus has emerged as a useful concept to describe and address the complex and interrelated nature of the global resource system, on which humans depend to achieve different social, economic and environmental goals [16]. In the present study, these links are evident and compromise the viability of the entire region.

The alternative visualized by the authors of this document implies a close collaboration between producers and the government. First of all, a gradual transition to the use of renewable energies is proposed, followed by a combination of power generation sources based on a model such as the following:

a) The agricultural producers must define, in a consensual manner, the percentage of energy that will be generated using renewable sources without disconnecting from the traditional power grid.
b) Individually, farmers with low-capacity desalination plants should try to comply with the percentage established by means of photovoltaic cells.
c) Large agricultural producers should partner with the government to establish marine wind farms in the area. These projects must be implemented gradually until they meet the established percentages. The timeframe also needs to be defined in a consensual manner between producers and government.
d) All producers must collaborate with research centers and universities in the region to find more efficient and cheaper ways to generate renewable energy in the medium term.

ACKNOWLEDGEMENTS

This research was supported by the Center for Innovation and Design in Engineering (CEID) of the Engineering College of CETYS University. The authors appreciate all its support. Also, the authors acknowledge the comments made by Dr. Josue López Leyva, Dean of the Renewable Energy Department of CETYS Ensenada Campus, to the initial manuscript.

REFERENCES

[1] Indicador Trimestral de la Actividad Económica Estatal. Instituto Nacional de Estadística y Geografía. Online. http://www.beta.inegi.org.mx/temas/itaee/. Accessed on: 22 Jan. 2019.
[2] Servicio de Información Agroalimentaria y Pesquera. Secretaría de Agricultura y Desarrollo Rural. www.gob.mx/siap. Accessed on: 19 Oct. 2018.
[3] Azuz-Adeath, I., González-Campos, C. & Cuevas-Corona, A., Predicting the temporal structure of the Atlantic Multidecadal Oscillation (AMO) for agriculture management in Mexico's coastal zone. *Journal of Coastal Research*, **35**(1), pp. 210–226, 2019.
[4] Resúmenes mensuales de temperaturas y lluvias. Servicio Meteorológico Nacional. https://smn.cna.gob.mx/es/climatologia/temperaturas-y-lluvias/resumenes-mensuales-de-temperaturas-y-lluvias. Accessed on: 16 Sep. 2018.
[5] Kay Bailey Hutchinson WTP. El Paso Water. www.epwater.org/cms/one.aspx?portalId=6843488&pageId=7422402. Accessed on: 22 Nov. 2018.
[6] Atlas Nacional de Zonas con Alto Potencial de Energías Limpias, Secretaría de Energía. https://dgel.energia.gob.mx/AZEL/. Accessed on: 4 Jun. 2018.
[7] Índice de Marginación por Localidad 2010, Consejo Nacional de Población. www.conapo.gob.mx/en/CONAPO/Indice_de_Marginacion_por_Localidad_2010. Accessed on: 19 Aug. 2018.
[8] Catálogo de SIG y Mapas en Línea, Instituto Municipal de Investigación y Planeación del Municipio de Ensenada. http://imipens.org/catalogo-sig-en-linea/. Accessed on: 23 Sep. 2018.
[9] Imágenes del Territorio. Instituto Nacional de Estadística y Geografía. www.inegi.org.mx/datos/?t=0130000000000000. Accessed on: 23 Sep. 2018.
[10] Fuentes-Yagüe, J.L. & García-Legaspi, G., *Técnicas de riego. Sistemas de riego en agricultura*. Grupo Mundi-Prensa: México, D.F., pp. 39–63, 1999.
[11] Estación Camalú, B.C., Variables Climáticas. Instituto Nacional de Investigaciones Forestales, Agrícolas y Pecuarias. http://clima.inifap.gob.mx/lnmysr/Estaciones/ConsultaDiarios15Min?Estado=2&Estacion=998421. Accessed on: 25 Jul. 2018.

[12] Allen, R.G., Pereira, L.S., Raes, D. & Smith, M., *Crop Evapotranspiration – Guidelines for Computing Crop Water Requirements*, FAO Irrigation and drainage paper 56, Rome, 1998. www.fao.org/3/X0490E/X0490E00.htm.
[13] Información por Cultivo Distrito de Ensenada, B.C. Sistema de Información Agroalimentaria de Consulta. www.gob.mx/siap/documentos/siacon-ng-161430. Accessed on: 8 Sep. 2018.
[14] *The European Offshore Wind Industry. Key Trends and Statistics 1st half of 2018*, Wind Europe: Brussels, pp. 1–14, 2018.
[15] Lundqvist, J., 2014 Stockholm Water Prize laureate's presentation: The power of water by Professor John Briscoe. *Aquatic Procedia,* **5**, pp. 3–7, 2015.
[16] Endo, A., Tsurita, I., Burnett, K. & Orencio, P.M., A review of the current state of research on the water, energy, and food nexus. *Journal of Hydrology: Regional Studies*, **11**, pp. 20–30, 2017.

FORMATE FISCHER–TROPSCH PROCESS FOR PRODUCING TRADITIONAL ENERGY CARRIERS WITH ZERO CARBON BALANCE

MIKHAIL GOTOVSKY, ALEXANDER GOTOVSKY, VITALY LYCHAKOV,
VLADIMIR MIKHAYLOV, YURY SUKHORUKOV & EKATERINA SUKHORUKOVA
Polzunov Institute, Russia

ABSTRACT

The Fischer–Tropsch process (F-T process) is known as a method of synthesizing organic products (including liquid fuel) from $CO + H_2$ mixture gases (synthesis gas) using various catalysts. The interest in the F-T process is associated with the global struggle against carbon emissions. Earlier we noted that the electrochemical synthesis of formic acid from water and atmospheric CO_2 can be a promising general scheme for obtaining "green fuel" with minus (negative) "carbonic effect" simultaneously "formate cycle" (synthesis of organic products from formic acid produced from atmospheric CO_2) fuel has zero carbon balance. This synthesis is brought to efficiency, surpassing natural photosynthesis. Synthesized formic acid can be used as a fuel for fuel cells and as a hydrogen accumulator. A semi-product of green fuel synthesis with zero carbon balance and synthetic organic chemical products with minus carbon balance (thanks to bonded carbon from atmospheric CO_2). From an economical point the direct use of formic acid would be more rational (with the exception of one stage), without first carbon monoxide separation. From the point of view of the physical chemistry, it is necessary to take into account the relatively low reactivity of formic acid in comparison with carbon monoxide. The traditional way of activation, by applying high temperatures, high pressures and specific catalysts in gas phase. However, using electromagnetic activation may be a more interesting pathway. This aspect will be discussed in this work. The final selection of the method based on the results of bench scale should be made on the basis of a complex criterion of the rate of passage of the modified F-T process and economic indicators: equipment cost, and energy consumption in terms of the mass of final products, taking into account of their cost. In this case, it appears an alternative way of obtaining some final products without using fossil hydrocarbons.

Keywords: Fischer–Tropsch process, formate cycle, green fuel, electromagnetic activation, zero carbon balance.

1 INTRODUCTION

The Fischer–Tropsch process (F-T process) has been known since the 1920s as a method of obtaining a number of organic products from a mixture of $CO + H_2$ gases produced by treating coal with water vapor. The mixture of $CO + H_2$ later was referred as synthesis gas. The same mixture can be received by non-oxidative pyrolysis of timber waste.

Depending on the ratio of the synthesis gas components, using different catalysts (usually transition metals from the series: iron, cobalt, nickel and copper), the following redox reactions take place:

$$CO + 2H_2 \rightarrow -CH_2- + H_2O,$$

or

$$2CO + H_2 \rightarrow -CH_2- + CO_2.$$

The symbol $-CH_2-$ is a mixture of organic products similar to alkane, resulting in each of the two reactions. The ratio of different organics depends on the physical conditions (temperature, pressure), on the catalyst used, and on the residence time in the reaction zone.

WIT Transactions on Ecology and the Environment, Vol 237, © 2019 WIT Press
www.witpress.com, ISSN 1743-3541 (on-line)
doi:10.2495/ESUS190141

During the Second World War, various versions of the F-T process were used in Germany and Japan to produce synthetic liquid fuels (ersatz gasoline), diene monomers of synthetic rubber (for car tires), as well as aromatics (hereafter processed into paints, varnishes and pharmaceuticals). In 1944, ersatz gasoline was produced in Germany in the amount of 124,000 barrels per day (6.5 million tons per year). In the post-war period, interest in the F-T process recurred during the 1973 oil crisis, but later disappeared when oil and natural gas prices fell.

In the 1980s, in Brazil, the F-T process was considered as one of the ways to produce synthetic liquid fuels from forest waste, but biotechnological approaches for processing into fuel alcohol (bioethanol) was recognized as more profitable.

Nowadays the interest in the F-T process is motivated by the problem of the global struggle with greenhouse gases, primarily with carbon emission. Carbon emission refers to the release of CO_2 from the combustion of hydrocarbon fuels derived from mineral raw materials: oil, natural gas, and coal. Owing to this struggle (the Kyoto Protocol, the Paris Agreement, etc.), the processes of producing "green fuel" with a "zero carbon-carbon balance" have become very relevant. All carbon in such fuels should be obtained by binding atmospheric CO_2, and subsequent combustion only returns the same CO_2 to the atmosphere. Also becoming relevant are industrial processes with a "minus carbon balance", i.e. those in which synthetic chemical products contain fixed carbon from atmospheric CO_2.

In [1], [2], it was noted that a promising general scheme for obtaining green fuel and products with a minus carbonic effect can be the "formate cycle". The basis of the formate cycle is the electrochemical production of formic acid from water and atmospheric CO_2:

$$CO_2 + 2H_2O + 2e^- \text{->} HCOOH + 2OH.$$

This process has been known since 1869, from the experiments of Nikolai Beketov from his electrochemical synthesis of sodium bicarbonate formates experiments. Much later, in the 2010s, due to the interest in "green energy", the process was brought (at lab scale) to a level even more energy-efficient than natural photosynthesis (that is, than the process of binding CO_2 in the form of carbohydrates to plants).

The resulting formic acid can be used in several ways, in particular:

1. As fuel for fuel cells;
2. As a hydrogen accumulator (hydrogen is released by the reaction HCOOH -> $CO_2 + H_2$);
3. As an intermediate for the synthesis of green fuel with zero (neutral) carbon balance, and synthetic organic chemical products with a minus carbon footprint.

With respect to the third point to catalytically convert formic acid into synthesis gas components necessary for the F-T process:

$$HCOOH \text{->} CO_2 + H_2,$$

$$HCOOH \text{->} H_2O + CO.$$

At the same time, the quality of the synthesis gas that should be obtained this way will be much higher than in the original process dating from the 1920s–1940s. There, synthesis gas was obtained from coal or from biological waste, and inevitably contained impurities of sulfur and nitrogen compounds, acting as catalytic poisons and preventing the selective conduct of the process, aimed at a number of pre-defined products.

In the case of the formate cycle, synthesis gas will generally not contain catalytic poisons and other components that reduce the selectivity of the F-T process, which will allow the use of a number of existing developments that require high purity at the inlet.

The disadvantage of implementing this approach is the need for three separate stages:

1. The release of hydrogen from formic acid;
2. Isolation of carbon monoxide from formic acid;
3. The actual F-T process.

A more rational approach would be the direct use of formic acid, without the preliminary isolation carbon monoxide.

At the level of chemical reactions, such an analogue of the F-T process would look like:

$$HCOOH + 2H_2 \text{->} \text{-}CH_2\text{-} + 2H_2O,$$

or

$$2HCOOH + H_2 \text{->} \text{-}CH_2\text{-} + CO_2 + 2H_2O.$$

The closest approach linked to this concept is the catalytic reduction (with H_2) of formic acid to methanol:

$$HCOOH + 2H_2 \text{->} CH_3OH + H_2O.$$

From a physical chemistry point of view the process under consideration, it is necessary to take into account the relatively low reactivity of formic acid in comparison with carbon monoxide. However there is a traditional way of activating it, which is by applying high temperatures, high pressures in gas phase, by designing specific catalysts.

However, an electromagnetic activation may be more interesting.

The theory of electromagnetic activation today is not sufficiently developed to allow calculating what kind of electromagnetic effect would lead to a satisfactory reaction rate, and which will be optimal from the point of view of energy consumption for the process.

Based on the chain free-radical version or the ion-radical version of the process, activation by an electrostatic field at the field strength near the physical boundary of the breakdown (occurrence of a corona gas discharge) may be promising [3].

Another option could be the exposure to microwave radiation with a wavelength of about 0.1 mm (i.e. in the overlap region with the infrared range, in which absorption peaks of the carbon–oxygen bond are possible with appropriate excitation of this bond and, accordingly, its activation) [4].

It should also be noted the possibility to combine electromagnetic and catalytic effects on the reaction medium, in particular, the use of the catalytic effect of the micro-needle surface, on which, when an electric potential is applied, many areas with high field strength or multiple corona discharge can be created. Even with a relatively small imposed electrical potential, there are a number of specific physicochemical effects associated with the local electric field strength around the microneedles, as noted in the work of Zaitsev in 2017 [5].

2 MILLER–UREY EXPERIMENTS

A significant amount of data on such organic synthesis reactions has been accumulated since 1953, which would be the experiments of Stanley Miller and Harold Urey [6]. These experiments were not planned as technologically oriented, but were carried out to test hypotheses of pre-biological evolution on Earth, which began about 4 billion years ago, in a reducing atmosphere containing hydrogen, methane, ammonia, carbon monoxide, and water vapor. The mixture of gases was exposed to a spark discharge simulating the periodic lightning in the atmosphere. Despite the low temperature and low intensity of electrical discharges, the degree of conversion to higher molecular weight organic compounds ranged from 10–15% after 170 hours.

Subsequently, the Miller–Urey experiments were reproduced under different conditions and showed that the yield of various organic compounds depends on the local electric fields at the surface of the solid phase.

Thus, with sufficient confidence it can be said that there is a mixed effect of the classical catalysis and electrical field initiation for a series of chemical reactions including the formation of new carbon–carbon bonds in the region of locally high field strength on a solid inorganic surface. This can serve as a starting point for the development of technological methods of a modified F-T process, based on a direct gas-phase reaction of formic acid with hydrogen, using a combination of inexpensive catalysts and an electric field.

Such methods theoretically open up the possibility not only for the reaction of formic acid with hydrogen, but for reactions with a third component: carbon dioxide. Under normal conditions, carbon in a CO_2 molecule exhibits low reactivity, in particular, due to the high symmetry of the molecule and zero dipole moment. However electromagnetic activation can lead to a change in the configuration of the electron orbitals, and the deviation of the molecule from a strictly linear structure. In this case, we can assume in the three-component system the following reactions:

$$HCOOH + CO_2 \text{ -> } HCOOH\text{-}COOH \text{ (oxalic acid formation)},$$

$$HOOC\text{-}COOH + H_2 \text{ -> } HOOC\text{-}CHO + H_2O \text{ (reduction to glyoxylic acid)}.$$

Glyoxylic acid is already quite high reactivity. Also, for any of the two types of electromagnetic activation discussed, it is possible to assume with high probability that the reaction proceeds in parallel:

$$CO_2 + H_2 \text{ -> } HCOOH.$$

Thus, in the above three-component system, an additional mole of CO_2 per mole of HCOOH can be involved in the F-T process, and an additional amount of hydrogen can be involved in binding CO_2, which greatly increases the efficiency of the proposed process for utilizing atmospheric carbon dioxide.

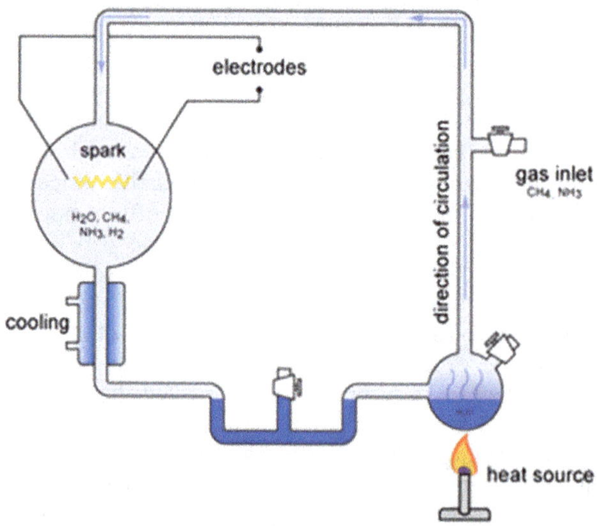

Figure 1: Scheme of Miller–Urey experiment.

3 USING DISPROPORTION (TRANSMUTATION) OF FORMIC ACID

Disproportionation, sometimes called dismutation, is a *redox reaction* in which a compound of intermediate oxidation state converts to two different compounds, one of higher and one of lower oxidation states. Although not widely accepted, disproportionation is sometimes used to describe any desymmetrizing reaction of the following type: 2 A → A' + A", regardless of any redox process. In other words disproportionation is a *chemical reaction*, typically a redox reaction, where a *molecule* is transformed into two or more dissimilar *products*. In a redox reaction, the species is simultaneously oxidized and reduced to form at least two different products.

Such exception would also be linked to the second preliminary stage which is the release of hydrogen from formic acid. Such an optimal analogue of the F-T process would involve a redox disproportionation of formic acid (dismutation and asymmetrization of the initial substance by hydrogen).

At of chemical reaction level, such a process could be presented such as:

$$3HCOOH \rightarrow -CH_2- + 2CO_2 + 2H_2O.$$

The most comparative process applied at lab scale is the disproportionation (dismutation) of benzoic acid or its salts in the presence of a catalyst:

$$C_6H_5COOH \rightarrow C_6H_6 + C_6H_4\,(COOH)_2.$$

It is also known reduction–oxidation disproportionation of carbon monoxide:

$$2CO \rightarrow C + CO_2,$$

and reduction–oxidation disproportionation of formaldehyde:

$$2HCHO + H_2O \rightarrow CH_3OH + HCOOH.$$

It should be noted that the second of the known paths leading to the Fischer–Tropsch reaction goes as follows:

$$2CO + H_2 \rightarrow -CH_2- + CO_2.$$

The latter partly involves also disproportionation.

Also partly disproportionation will be the reaction of formic acid and hydrogen:

$$2HCOOH + H_2 \rightarrow -CH_2- + CO_2 + 2H_2O.$$

Exploring the possible ways to disproportionate formic acid (or formates) is a nontrivial task, but the result looks very promising in an economic sense in two obvious ways. First way is the decrease of number of process stages. Second way is absence of costly catalysts. Both ways are typical for cost reduction in chemical technology. The one-step production of synthetic green fuel from formic acid, which is almost identical to gasoline, will allow, without additional changes, to transfer classic motor transport to a fuel scheme with zero carbon balance. This will minimize the costs of implementing measures related to the Kyoto Protocol and the Paris Agreement.

4 PROPOSALS FOR DEVELOPMENT ELECTROMAGNETIC ACTIVATION

The practical steps to develop a new formate based F-T process can be performed at a laboratory scale to identify possible reaction paths. In particular, it can be achieved by verification of versions of effective disproportionation in a strong electric and/or electromagnetic field.

In this regard, there is an analogy with the above-mentioned possibilities of electromagnetic initiation of a series of reactions of formic acid with hydrogen, as well as with carbon dioxide, but with a wider variety of possible methods. Overall, three broad classes of methods of activation of the electric (electromagnetic) field could be considered:

1. Activation in a homogeneous gaseous environment: this involves an activation in the corona gas discharge of an electrostatic field, or an activation by microwave radiation with a wavelength of about 0.1 mm (mentioned above);
2. Activation in a two-phase medium with two continuous phases or with a dispersed liquid phase: this is for example an activation by a glow discharge in the gas phase in contact with the liquid phase (in particular, with aerosol droplets);
3. Activation in a non-stationary two-phase medium with a dispersed gas phase: this is the activation method in a fluid when applying a high power ultrasonic field. When this occurs, a phenomenon of cavitation with luminescence effects upon the collapse of microbubbles (sonoluminescence).

Figure 2: Corona discharge photography.

The mechanism of sonoluminescence is still insufficiently studied, but the excitation of interatomic bonds is undoubted, since only these effects can explain the high-brightness light flashes [7].

"Sonochemistry is studied primarily by chemists and sonoluminescence mainly by physicists, but a single physical phenomenon – acoustic cavitation – unites the two areas. The physics of cavitation bubble collapse is relatively well understood by acoustical physicists but remains practically unknown to chemists. By contrast, the chemistry that gives rise to electromagnetic emissions and the acceleration of chemical reactions is familiar to chemists, but practically unknown to acoustical physicists" [8].

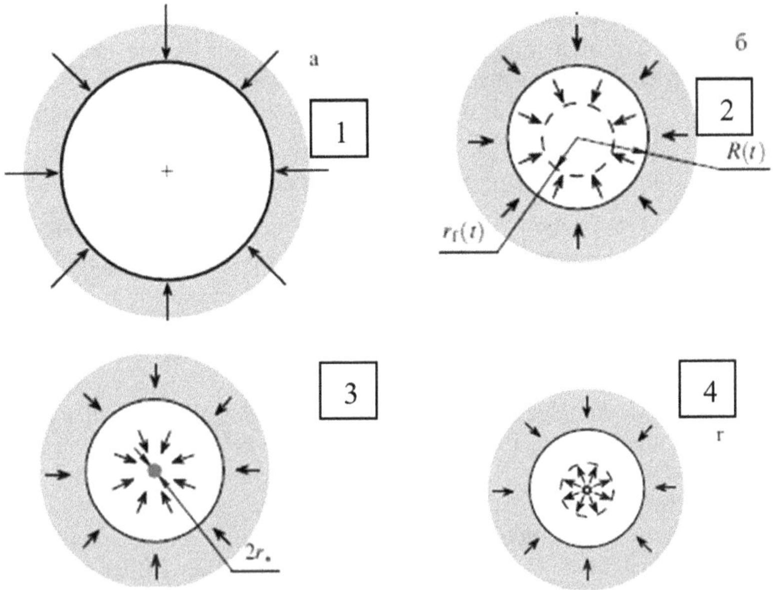

Figure 3: Scheme of sonoluminescence [9]. 1: Compression start; 2: Shock wave formation; 3: Extremal confined plasma; 4: Expansion of the compression zone. R, r = bubble and shock wave radii.

Here it will be useful to mention the great contribution to the study of this process, made by Nigmatulin et al. [9]. Although the main goal of their research was to prove the possibility of the emergence of local thermonuclear fusion under these conditions, they succeeded in significantly increasing the level of description of the physics of sonoluminescence processes.

5 CONCLUSIONS

Initiation of the modified F-T process in electromagnetic field allows to organize more technological regimes of producing final products (hydrocarbons) from formic acid bypassing intermediate stages of classic F-T process. In particular high technological value of class 2 and class 3 processes should be noted. And problem of hydrocarbons separation from formic acid and water is easily solved because of hydrocarbons insolubility.

As mentioned above, the final choice of the method based on the results of bench studies should be made on the basis of a complex criterion of the rate of passage of the modified F-T process, the degree of conversion of the starting materials, and economic indicators. The latter involves (but are not limited to) equipment cost, and energy consumption in terms of the mass of the final products, taking into account their cost. In this case, an alternative way of obtaining such end products without using fossil hydrocarbons can play a special role.

Installations facilities with the modified F-T process may be of particular economic importance for regions that don't possess their own sources of hydrocarbons, but that have an excess of electricity produced (due to the presence of nuclear, solar, wind, geothermal, hydropower, or tidal power plants). In such a region, the installation of modified F-T process can, due to atmospheric carbon dioxide, solve the problem of providing its own raw materials for the organic chemical industry, with its own authentic gasoline or diesel (but with zero carbon balance), as well as fuel for transport on formate or hydrogen fuel cells. In the case of a liquid-phase class of a modified F-T process, the installation can be sufficiently environmentally friendly to be placed on sites without exclusion zone from natural, agricultural, or residential facilities, which is significant for regions with high population density.

REFERENCES

[1] Gotovsky, M., Gotovsky, A., Mikhailov, V., Kolpakov, C., Lychakov, V. & Sukhorukov, Y., Formic acid cycle as partial alternative to Allam cycle less expensive and more simple. *TI-IJES*, **61+1**(2), p. 49, 2018,

[2] Gotovsky, M., Gotovsky, A., Mikhailov, V., Lychakov, V., Sukhorukov, Y. & Sukhorukova, E., Formate cycle: The third way in green energy. *ICEBE 2019*, Singapore, p. 29, 2019.

[3] Alexankin, M., Monomolecular reactions of organic radical cations in a strong electric field, Doctoral thesis, Kiev, 1984.

[4] Mikheev, A. & Pankrushina, N. Peculiarities and possibilities of microwave chemistry. *Science in Syberia*, **21**, p. 9, 2010. (In Russian.)

[5] Zaitsev, S., *Microostria, Application, Manufacturing*, 2017. (In Russian.)

[6] Bada, J., New insights into prebiotic chemistry from Stanley Miller's spark discharge experiments. *Chemical Society Reviews*, **42**(5), pp. 2186–2196, 2013.

[7] Margulis, M.A., *Sonochemistry and Cavitation*, Langhorne, 1995.

[8] Apfel, R. et al. (eds), *Sonochemistry and Sonoluminescence*, 1999.

[9] Nigmatulin, R. et al., On thermonuclear processes in cavitation bubbles. *Uspekhi Fizicheskikh Nauk*, **184**(9), pp. 947–960, 2014. (In Russian.)

DISTANCE MEASURES FROM REPLICATOR SYSTEMS WITH NONLINEAR PAIRWISE INTERACTIONS FOR ENVIRONMENTAL RISK MANAGEMENT

YURI A. PYKH[1] & IRINA G. MALKINA-PYKH[2]
[1]St. Petersburg Mathematical Society, Russia
[2]St. Petersburg State Institute of Psychology and Social Work, Russia

ABSTRACT

Replicator dynamics is an evolutionary strategy well established in different disciplines of environmental sciences. It describes the evolution of self-reproducing entities called replicators in various independent models of, for example, genetics, ecology, prebiotic evolution, and socio-biology. So, the replicator systems arise in an extraordinary variety of ecosystem modelling situations. In this report we combine the three universalisms in modelling: nonlinear pairwise interactions concept, dynamical systems theory and generalized relative entropy as a distance measure for environmental risk management. We introduce two hypotheses concerning the structure and types of properties of the system's entities' nonlinear interactions and derive generalized replicator dynamic equations. If there exists a nontrivial equilibrium point for a generalized replicator system then we construct an entropy-like Lyapunov-Meyer function (LMF) for this system and prove that it is a relative entropy function or the function of information divergence. We have proven that negative relative entropy is a convex function for a probability space and receive new distance measures (divergence) between two probability distributions. In conclusion, we show, as an example, that relative entropy as distance measures may be used as sustainability indicators for estimating the efficiency energy use of wastewater treatment. These results allow us to suggest that these new distance measures can be applicable to a wide set of real environmental situations.

Keywords: nonlinear pairwise interactions, replicator dynamics, Lyapunov functions, distance measure, divergence, sustainable development.

1 INTRODUCTION

Sustainable development is a complicated concept that has arisen to synthesize ideas that simultaneously address the risk of environmental impacts and the needs of people. Sustainable development commonly is defined as the process of meeting the needs of the present generation without compromising the ability of future generations to meet their needs. This concept arose in the late 1980s as an approach to balance economics and environment. Sustainable development embodies several advancements over other models that have characterized the development process. One advancement is drawing distinctions between the growth component and the efficiency component of development. It is evident that measures of divergence between two points play a key role in many environmental problems. One can find descriptions of energy, water and agricultural aspects of sustainable development in the books [1]–[3].

Mathematical models are useful tools for both understanding the dynamics of populations and impact assessment for several reasons. In the first place, models can lead to an increase of the general understanding of a system and help generate hypotheses and direct field experiments. In the second place, models can be used as a tool for evaluating alternative plans (scenarios) or by providing rules of thumb to policy makers.

Ecosystem stability and the response of ecosystems to disturbance are of crucial importance for conservation management. A variety of ecological interpretations have been given to the term "stability". The most generally used concept refers to the stability in the

WIT Transactions on Ecology and the Environment, Vol 237, © 2019 WIT Press
www.witpress.com, ISSN 1743-3541 (on-line)
doi:10.2495/ESUS190151

vicinity of an equilibrium point in a deterministic system. Pimm [4] distinguished two aspects of this stability: "resilience", the speed at which a system returns to its equilibrium following a perturbation, and "resistance", the ability of a system to withstand displacement by a disturbance in its environment. These two properties are usually held to be responsible for the persistence of ecosystems in an unpredictable environment. A "stable" ecosystem might be expected to exhibit both of these properties to some degree. The linkages between resilience and the stability of dynamical systems are discussed, along with its role in understanding of the evolution of such systems.

Many problems of investigating stability and calculating characteristics of dynamical systems (time and value of transient processes, speed of convergence, stability region, integral criteria, etc.) are solved using the direct Lyapunov method. The method is based on searching the function having certain properties along the system solutions. See, for example Common and Perrings [5].

The primary goal of this paper is to study the implications of Lyapunov functions theory for the analysis of nonlinear pairwise interactions replicator systems. We build Lyapunov-Meyer functions as entropy-like functions (distance measures). The existence of the Lyapunov-Meyer functions allows us to construct practically all known entropy and relative-entropy functions. Our investigations are based mainly on the stability postulate stated by Chetaev in 1936 [6].

Stability postulate. Stability, which is a fundamentally general phenomenon, apparently, must manifest itself in basic laws of nature in some way. If knowledge is constructed from the requirement of small deviations from nature, then scientific thinking must (or can) rely on some Lyapunov function. Certainly, this function always exists according to the stability postulate.

Many years later in 1967 stability postulate, in more narrow sense, was used by Coleman and Mizel in their famous paper "Existence of Entropy as a Consequence of Asymptotic stability" [7].

In 1968, this postulate was stated mathematically by Meyer [8], who proved that, for dynamical systems whose limit sets consist of only isolated rest points or cycles, i.e., for Morse–Smale systems, there always exists a Lyapunov function, which increases on the set of wandering points of the system. Meyer, followed by Smale [9], suggested the term energy functions for such functions; however, it is more natural and convenient to refer to them as Lyapunov–Meyer functions. One of the best-known examples of such function is entropy in Boltzmann's H-theorem.

Prigogine [10] first pointed out the importance of the relationship between Lyapunov functions and entropy: "The positive time direction is associated with the increase of entropy. Let us emphasize the strong and very specific way in which the one-sidedness of time appears in the second law. According to its formulation it leads to the existence of a function having quite specific properties as expressed by the fact that for an isolated system it can only increase in time. Such function plays an important role in modern theory of stability as initiated by the classic work of Lyapunov. For this reason they are called Lyapunov functions (or functional).

The entropy S is a Lyapunov function for isolated systems. As shown in all textbooks, thermodynamic potentials such as the Helmholtz or Gibbs free energy are also Lyapunov functions for other "boundary conditions" (such as imposed values of temperature and volume). In all these cases the system evolves to an equilibrium state characterized by the existence of a thermodynamic potential. This equilibrium state is an "attractor" for non-equilibrium states.

For dynamical systems arising from physics the Lyapunov functions will typically have a thermodynamic interpretation (energy, entropy etc.) but its origin is not evident.

In an attempt to show how such an interpretation this idea might proceed, we analyze below a class of macrosystems with nonlinear pairwise interactions for with an assumption of structure and type of interactions has consequences that there exist entropy-like functions.

2 EQUATIONS OF MACROSYSTEMS WITH NONLINEAR PAIRWISE INTERACTIONS

One of the important assumptions in classic Lotka-Volterra and replicator systems is that the functional responses are linear functional responses. Usually additive linear models for predicting combined interactions effects cannot account for non-linearities in combined functional response introduced by non-trophic interactions Ayala et al. [11]. Exponential and power law responses were used to model social dynamics by Beechman and Farnsworth [12]. The PhD thesis by Merrifield [13] provides a review on the concept of a force including nonlinear in different interacting systems in Chapter 5.

Consider a macrosystem formed by a sufficiently large number *N* of interacting objects. The classical definition of macrosystems is as follows: these are systems in which a chaotic behavior at the microlevel transforms into a deterministic behavior at the macrolevel. Suppose that, at a moment *t*, the macrosystem under consideration contains *n* different types of objects and the number of objects of type i is $x_i(t)$, where $i = 1,...,n$ and $\sum_{i=1}^{n} x_i(t) = N(t)$. Consider the relative numbers $p_i(t) = x_i(t)/N(t)$, $p_i(t)$ of various types' objects. Obviously, $\sum_{i=1}^{n} p_i(t) = 1$, i.e., vector $\mathrm{p}(t) \in \sigma_p^n = \{\mathrm{p} \in \mathbb{R}^n : p_i \geq 0, i = 1,2,...,n, \mathbf{e}^T\mathrm{p} = 1\}$, where σ_p^n is the standard simplex in Euclidean *n*-space $\mathbb{R}^n$ and, $\mathbf{e}$ is the vector of ones. Thus, the state of such a macrosystem at each moment t *is* determined by the vector $\mathrm{p}(t) = (p_i(t),..., p_n(t))$.

We consider the systems in which only pairwise interactions occur during a short time interval $(t, t + \Delta t)$; in other words, those systems in which simultaneous interactions of more than two particles are impossible.

We make the following two assumptions about interactions between objects in a macrosystem [14].

Hypothesis 1: The interaction between objects of types i and j is characterized by the so-called interaction strength, which we denote by w_{ij} and regard as a quantitative characteristic of the effect of the interaction between two objects of types i and j on the rate of change of the relative number $p_i(t)$ of objects of type i.

Remark 1: Apparently, the notion of an interaction strength first appeared in mathematical ecology. A fairly detailed study of this notion and a survey of related results are contained in Ayala et al. [11].

Remark 2: Note that the definition given above does not imply that $w_{ij} = w_{ji}$. We also emphasize that the asymmetry of results of interaction plays an essential role in many cases [15].

Hypothesis 2: For each macrosystem under consideration, there exists a set of probability distribution functions $f_i(p_i)$, where $i = 1,...,n$, which determine the probability of the interaction of each object of type i with any other object in the macrosystems. Thus, the probability of the pairwise interaction between objects of types i and j is determined by the product $f_i(p_i)f_i(p_i)$.

Remark 3: The probability distribution functions f_i (nonlinear response functions) obeys the conditions $f_i(0)=0$, $\partial f_i/\partial p_i > 0$ for $p_i > 0$, $\partial f_i/\partial p_i \geq 0$ for $p_i = 0$ and $f_i(1)=1$ [16].

Note that the standardization condition for response functions was investigated in the article by Pykh and Malkina-Pykh [17].

It follows from the two hypotheses stated above that, for small Δt , the dynamics of the system is determined by the relation:

$$p_i(t+\Delta t) = p_i(t) + \Delta t \sum_{j=1}^{n} w_{ij} f_i(p_i(t)) f_j(p_j(t)) + o(\Delta t) \ ,$$

subject to the constraints $\sum_{i=1}^{n} p_i(t) - 1 = 0$, and $f_i(0)=0$, for $i=1,2,\ldots,n$ which ensure the invariance of the simplex σ_p^n and all of its faces. Using an approach common in analytic mechanics, we treat the system under consideration as a nonfree system subject to the ideal holonomic constraints:

$$f_i(p_i)\left(\sum_{i=1}^{n} p_i(t) - 1\right) = 0, \quad i=1,\ldots,n$$

and apply the principle of replacing constraints by reaction forces to this system [18]. Passing to the limit as $\Delta t \to 0$ and taking into account the obvious identity $\sum \dot{p}_i = 0$ for the sum of derivatives, we obtain the following equation for the constraint multipliers (the indeterminate Lagrange multipliers) λ_i :

$$\sum_{ij}^{n} w_{ij} f_i(p_i) f_j(p_j) - \sum_{i=1}^{n} \lambda_i f_i(p_i) = 0$$

This equation has two obvious solutions, the trivial solution $\lambda_i = \sum_{j=1}^{n} w_{ij} f_j(p_j)$ and the nontrivial solution:

$$\lambda_i = \lambda = \frac{\sum_{ij}^{n} w_{ij} f_i(p_i) f_j(p_j)}{\sum_{i=1}^{n} f_i(p_i)}, \quad i=1,\ldots,n$$

Using the nontrivial solution and denoting the sum $\sum_{i=1}^{n} f_i(p_i)$ by $\theta(\mathrm{p})$, we obtain the following system of differential equations determining the evolution of the probability distribution $\mathrm{p}(t)$:

$$\dot{p}_i = f_i(p_i)\left(\sum_{j=1}^{n} w_{ij} f_j(p_j) - \theta^{-1}(\mathrm{p}) \sum_{ij}^{n} w_{ij} f_i(p_i) f_j(p_j)\right), \quad i=1,\ldots,n . \qquad (1)$$

This system of equations was first obtained from balance considerations in Pykh [19] without the use of the Lagrange method. In what follows, it is convenient to pass to the matrix form. System (1) is written in this form as:

$$\dot{\mathrm{p}} = D(\mathrm{f})\left(\mathrm{Wf} - \mathbf{e}\theta^{-1}(\mathrm{p})\langle \mathrm{f}, \mathrm{Wf}\rangle\right). \tag{2}$$

Here, $\mathrm{f}(\mathrm{p})$ is the vector $\left(f_1(p_1),\dots,f_n(p_n)\right)$; $D(\mathrm{f}) = diag(f_1, f_2 \dots, f_n)$; $\mathrm{W} = \left(w_{ij}\right)$ is the matrix of interactions; and $\theta(\mathrm{p}) = \langle \mathbf{e}, \mathrm{f}(\mathrm{p})\rangle$, where $\langle \cdot, \cdot \rangle$ denotes inner product. Obviously, since $\langle \dot{\mathrm{p}}(t), \mathbf{e}\rangle \equiv 0$ and $f_i(0) = 0$ it follows that the simplex σ_p^n and each of its faces are invariant sets for system (2). Note that system (2) determines the dynamics of objects with nonlinear pairwise interactions; the matrix W determines the structure of interactions, and the response functions determine their types. The elements of the matrix W are generalized interaction strengths, i.e., by analogy with the thermodynamics of irreversible processes, are "reasons" causing changes in the speed of flows.

3 ESCORT DISTRIBUTIONS

Let us rewrite eqn (2) as:

$$\dot{\mathrm{p}} = \theta D(\mathrm{f})\left(\mathrm{Wf}\theta^{-1} - \mathbf{e}\mathrm{E}(\mathrm{p})\right), \tag{3}$$

where $\mathrm{E}(\mathrm{p}) = \theta^{-1}(\mathrm{p})\langle \mathrm{f}, \mathrm{Wf}\rangle$. Using the terminology of the theory of neural networks, we refer to $\mathrm{E}(\mathrm{p})$ as the energy function of the macrosystem under consideration. Note that in the population genetic theory this function named as population fitness [20], and in the evolutionary game theory as utility function [21]. System (3) and the energy function $\mathrm{E}(\mathrm{p})$ naturally determine the introduction of new additional variables:

$$x_i(\mathrm{p}) = f_i(p_i)\theta^{-1}(\mathrm{p}) \qquad i = 1,\dots,n. \tag{4}$$

Obviously, $\mathrm{x} = (x_1, x_2 \dots, x_n) \in \sigma_x^n = \left\{\mathrm{x} \in \mathbb{R}^n : x_i \geq 0, \mathbf{e}^{\mathrm{T}}\mathrm{x} = 1\right\}$ for $\mathrm{p} \in \sigma_p^n$. The indices x and p are used in the notation of simplexes in order to avoid confusion. Consider change (4) in more detail. If this is a diffeo-morphism, then it can be regarded not only as a simplifying change of variables customary in the theory of differential equations but also as the definition of a set of quantities with particular physical meaning.

Theorem 1. [22]: Under the conditions (remark 3) used for the response functions f_i, for $\mathrm{p} \in \sigma_p^n$, there exists a one-to-one inverse mapping to (4), which is defined by:

$$p_i = f_i^{-1}(x_i) \Big/ \sum_{j=1}^{n} f_j^{-1}(x_j), \quad i = 1,\dots,n, \tag{5}$$

where $f_i^{-1}(\cdot)$ denotes the function inverse to $f_i(\cdot)$.

Remark 4: Recall that the notation $f_i^{-1}(\cdot)$ is used for both functions inverse in the sense of function theory and functions inverse in the algebraic sense. It is always clear from the context what is meant.

4 EQUILIBRIUMS

To go further, we need the following assertion [22]:

Statement 1: System (1) is invariant with respect to the replacement of the interaction matrix W by a perturbed matrix $W_\zeta = \left(W + e\zeta^T(p)\right)$, where the components of the vector function $\zeta(p) = (\zeta_1(p),\ldots,\zeta_n(p)) : \sigma_p^n \to \mathbb{R}^n$ are bounded on σ_p^n.

We proceed to consider questions related to the existence of a nontrivial equilibrium point $\hat{p} \in \mathrm{Int}\,\sigma_p^n$ for system (1). In Pykh [23], it was proved that $\hat{p} > 0$ exists if and only if all components of the vector $W^{-1}e$ are of the same sign. It is clear already from this result that properties of the inverse interaction matrix play a substantial role in the evolution of macrosystems.

Statement 1 implies that the coordinates of a nontrivial equilibrium point remain invariable $\forall\zeta \in \mathbb{R}^n$. So we have:

Statement 2. [22]: If system (4) has a nontrivial equilibrium point $\hat{p} \in \mathrm{Int}\,\sigma_p^n$, then:

$$\frac{\hat{f}}{\langle \hat{f}, e\rangle} = \frac{W^{-1}e}{\langle e, W^{-1}e\rangle} = \frac{W_\zeta^{-1}e}{\langle e, W_\zeta^{-1}e\rangle} = \hat{x} \qquad \forall\zeta(p) \in \mathbb{R}^n. \tag{6}$$

5 ENTROPY-LIKE LYAPUNOV-MEYER FUNCTIONS

Theorem 2. [22]: If system (1) has a nontrivial equilibrium point $\hat{p} \in \mathrm{Int}\,\sigma_p^n$ and the matrix $\left(W_\zeta^T + W_\zeta\right)$, $\forall\zeta \in \mathbb{R}^n$ has $n-1$ negative characteristic numbers, then the function:

$$H(p) = \sum_{i=1}^{n} \int_{\hat{p}_i}^{p_i} \frac{\hat{f}_i dx}{f_i(x)} + C, \tag{7}$$

where C is a constant, is a Lyapunov-Meyer function for system (1) in $\mathrm{Int}\,\sigma_p^n$, and the energy function $E(p)$ of the system attains its maximum value $E(\hat{p})$ as $t \to \infty$.

It is obvious that function $H(p)$ depends from interaction matrix W only through $\hat{p}$ and its characteristic numbers. It means that for each $H(p)$ we have a set of different systems (1) and can use this function for any two points $p(t)$, $\hat{p} \in \mathrm{Int}\,\sigma_p^n$.

In the next section we use Theorem 2 to construct entropy characteristics and "distances" between probability characteristics. First, we note that, if $G : \mathbb{R}^1 \to \mathbb{R}^1$ is a monotonically increasing smooth function and $C_1 > 0$, $C_2 > 0$ and C_3 are the numbers, then the function:

$$H_G = C_1 G\left(C_2 H(p)\right) + C_3$$

is also a Lyapunov-Meyer function for system (1).

6 DISTANCE MEASURES

Following Shannon information measure let us introduce $S(\mathrm{p})$ as negative entropy defined by:

$$S(\mathrm{p}) = -H(\mathrm{p}) = -\sum_{i=1}^{n}\int_{\hat{p}_i}^{p_i} \frac{\hat{f}_i dx}{f_i(x)}, \qquad p \in \sigma_p^n.$$

It is easy to see, that:

$$\lambda \sum_{i=1}^{n}\int_{\hat{p}_i}^{p_i} dx = \lambda \sum_{i=1}^{n}(p_i - \hat{p}_i) \equiv 0 \quad \forall p, \hat{p} \in \sigma_p^n$$

and we clearly have the following relationship:

$$S(\mathrm{p}) = -\sum_{i=1}^{n}\int_{\hat{p}_i}^{p_i}\left(\frac{\hat{f}_i dx}{f_i(x)} + \lambda\right)dx \quad \forall \lambda \in \mathbb{R}^1.$$

Since, by the evident equality $\partial S(\mathrm{p})/\partial p_i = 0$, with $p = \hat{p}$, we have that $\lambda = -1$. It is clear, that λ is a type of Lagrange multiplier. Therefore:

$$S(\mathrm{p}) = \sum_{i=1}^{n}\int_{\hat{p}_i}^{p_i}\left(1 - \frac{\hat{f}_i}{f_i(x)}\right)dx \qquad p \in \sigma_p^n. \tag{8}$$

The gradient of $S(p)$ with $p \in \mathrm{Int}\sigma_p^n$ is given by partial derivatives:

$$\nabla S(\mathrm{p}) = \left(1 - \frac{\hat{f}_i}{f_i(p_i)}\right)$$

that is strictly monotonically increasing for each p_i and Hessian of $S(p)$ on $\mathrm{Int}\sigma_p^n$ is $\mathcal{H}(\mathrm{p}) = diag\left(\hat{f}_i \frac{\partial f_i}{\partial p_i} f_i^{-2}(p_i)\right)$. Since $\partial f_i/\partial p_i > 0$ it follows that the Hessian is positive definite, and the function $S(p)$ according to definition is convex on $\mathrm{Int}\sigma_p^n$.

Another very well-known definition of convexity Jensen [24], Hardy et al. [25] is the next inequality with $p, \mathrm{q} = \mathrm{Int}\sigma_p^n$:

$$S(\mathrm{p}) - S(\mathrm{q}) - \langle \nabla_q S(\mathrm{q}), \mathrm{p} - \mathrm{q} \rangle \geq 0. \tag{9}$$

Evident that the expression from left-side inequality is the Bregman divergences denote by $B_s(\mathrm{p,q})$. This name was first given the by Censor and Lent [26]. Bregman divergence or Bregman distance [27] is similar to a metric but does not satisfy the triangle inequality or symmetry. Using inequality (9) we can receive new weighted distance measure between two probability distributions:

$$B_s(\mathrm{p,q}) = \sum_{i=1}^{n}\left(\int_{p_i}^{q_i}\frac{\hat{f}_i dx}{f_i(x)} + \frac{\hat{f}_i}{f_i(q_i)}(p_i - q_i)\right) \geq 0$$

It is obvious that $B_s(\mathrm{p,q})$ for p=q , and $B_s(\mathrm{p,\hat{p}}){=}S(p)$.

7 LEGENDRE-DONKIN-FENCHEL TRANSFORMATIONS

The Legendre-Donkin-Fenchel (LDF) transformation is a mathematical concept of great significance to thermodynamics, mechanics and probability theory is named after Legendre [28], Donkin [29] and Fenchel [30].

Let us denote $y_i = \partial S(\mathrm{p})/\partial p_i$. Since y_i is strictly monotonically increasing for p_i , this makes it clear that y_i and p_i are mutually reciprocal. The transformation from p to y is called LDF-transformation. We can find a convex function Y of y defined by:

$$\Upsilon(y) = \sum p_i y_i - S(p). \tag{10}$$

when y and p are respective coordinates of the same point, and the inverse transformation from y to p is given by the gradient:

$$p = \nabla \Upsilon(y)$$

Concerning the (8) we clearly have:

$$y_i = \frac{\partial S(p)}{\partial p_i} = 1 - \frac{\hat{f}_i}{f_i(p_i)}$$

and:

$$p_i = f_i^{-1}\left(\frac{\hat{f}_i}{1-y_i}\right) = \frac{\partial Y}{\partial y_i}$$

for $y_i \in \left(-\infty, 1-\hat{f}_i\right)$. We also denote $\partial Y/\partial y_i = \varphi(y_i)$ and after simple counting we have the following relationship:

$$\Upsilon(y) = \sum_{i=1}^{n}\left(y_i \varphi(y_i) + \int_{0}^{\varphi(y_i)}\left(\frac{\hat{f}_i}{f_i} - 1\right)dx \right)$$

After simple computation one can check that $\partial Y / \partial y_i = p_i$. Another form of LDF-transformation, together with (10) is:

$$\Upsilon(y) = \max_{p}\left(\langle p, y\rangle - S(p)\right)$$

Follow Amari [31] and definition (10) one can define a divergence function (distance measure) between two points p and q on σ_p^n in the dual coordinates. This divergence function will be similar to the squared Riemannian distance. Additional new information on LDF-transformation is presented in Nielsen [32] and Zia et al. [33] papers.

8 ENTROPY CHARACTERISTICS AND DISTANCE FUNCTIONS

Note that Theorem 1 can be used in two ways:

1. The response functions can be found for previously known entropy characteristics. Relevant examples for the Shannon and Tsallis entropy were given in Pykh [34]. It is easy to show that this approach yields response functions for all the entropy characteristics proposed in Esteban and Morales [35].
2. New entropy characteristic can be derived from some functions obeying the conditions stated for response functions. As an example, we consider the logistic function, which is widely applied in mathematical ecology and economics. This function is defined as follows:

$$f(x)=\frac{c}{(b+c)}\frac{(1-e^{-\alpha x})}{(b+ce^{-\alpha x})},$$

where $b>0$, $c>0$, $\alpha>0$.

It is obvious that $f(0)=0$, $f(+\infty)=c/b(b+c)$. Evaluating the corresponding integral and taking into account that arbitrary parameters can be picked in (8) (to get rid of the factors and summands), we derive the following expression for the logistic entropy $H_l(p)$:

$$H_l(p)=\sum_{i=1}^{n}\ln(1-e^{-\alpha p_i}).$$

Recall that $H_l(p)$ is a Lyapunov energy function for system (1) if and only if the conditions of Theorem 2 are satisfied and the equilibrium $\hat{p}$ is a uniform distribution; i.e. $\hat{p}=en^{-1}$. Since all the response functions are identical in the case under study, it is necessary and sufficient that W be quasi-stochastic. Note that the interest in objects described by noncanonical distributions has increased noticeably over resent years [35]. It is this type of distribution that can be derived from formula (8) suggested for entropy characteristics. For further use of these characteristics, it is natural to apply Jaynes' principle of maximum information entropy [36].

Now, we underline that distance functions are defined in the case where the equilibrium distribution in system (1) is not uniform. As in the previous case, we have that new distance functions between distributions can be derived from given response functions. As before, we consider the logistic function as an example. Let the conditions of Theorem 2 be satisfied. Evaluating the integral in (8) gives the function

$$H_l(p,\hat{p})=\sum_{i=1}^{n}\hat{f}_i\left(\frac{1-e^{-\alpha p_i}}{1-e^{-\alpha\hat{p}_i}}\right), \tag{11}$$

that, by analogy, is naturally called the relative logistic entropy.

9 ESTIMATIING OF WASTEWATE TREATMENT (EXAMPLE)

Sustainable development is one of the great hopes for conservation and proper use of resources including biodiversity. Instead of a focus on biodiversity itself, ecosystems sustainability represents a focus on a set of ecosystems properties, especially on ecosystems resilience, persistence and integrity [37]. One of the most important characteristics among them is the measures of divergence between two points that plays a key role in many ecosystems' problems. As it was shown in the article of Sobczuk and Lagod on the base of the results from [37]–[38] for Wastewater Treatment Plant (WWTP) [39]: "the results of bioindication, apart from the other physical and chemical parameters, may be used in wastewater plant treatment process management". Note that authors mean there "sustainable management".

In Sobczuk and Lagod [39], the comparison of different distance measures presented in Fig. 1 demonstrates that function (11) is the best distance measure of all applied methods for estimating the efficiency of wastewater treatment. Also in Sobczuk and Lagod [39], the authors pointed out that entropy and distance measures should be appropriately chosen for complex data sets comparison. Eqn (1) with accompanying theory allows to generate distance measures according to data set character. It shows up that the distance measure (11) is best fitted to recognize differences in microbial distribution functions described there.

10 CONCLUSIONS

We note that the approach suggested for constructing entropy characteristics and divergences between probability distributions based on the direct Lyapunov method applied to kinetic replicator equations (i.e., equations determining the dynamics of systems with nonlinear pair interactions) provides new strategies in the macroscopic analysis of complex open objects in various areas of the natural sciences. It should also be emphasized that this approach makes it possible to introduce mixed entropy, i.e., an entropy characteristic arising when the interacting subsystems have different response functions. Due to the possibility to generate measures according to the functional form of the response function in (1) one can find the best distance measure for the certain purpose. More information about the systems with pairwise interactions can be obtained in the preprint [40].

REFERENCES

[1] Malkina-Pykh, I.G. & Pykh, Y.A., *Sustainable Energy: Resources, Technology and Planning,* Computational Mechanics Publications: Southampton and Boston, 2002.

[2] Malkina-Pykh, I.G. & Pykh, Y.A., *Sustainable Water Resources Management,* Computational Mechanics Publications: Southampton and Boston, 2002.

[3] Malkina-Pykh, I.G. & Pykh, Y.A., *Sustainable Food and Agriculture,* Computational Mechanics Publications: Southampton and Boston, 2003.

[4] Pimm, S.L., The complexity and stability of ecosystems. *Nature,* **307**, pp. 321–326, 1994.

[5] Common, M. & Perrings, C., Towards an ecological economics of sustainability. *Ecological Economics,* **6**, pp. 7–34, 1992.

[6] Chetaev, N.G., Stability and classical physical law. *Proceedings of Kazan Aviation Institute,* **6,** pp. 3–5, 1936.

[7] Coleman, B.D. & Mizel, V.J., Existence of entropy as a consequence of anymtotic stability. *Archive for Rational Mechanics and Analysis,* **25**, pp. 243–270, 1967.

[8] Meyer, K., Energy functions for Morse-Smale systems. *American Journal of Mathematics,* **90**, pp. 1031–1040, 1968.

[9] Smale, S., On Gradient dynamical systems. *Annals of Mathematics*, **74**, pp. 199–206, 1961.

[10] Prigogine, I., *Time, structure and fluctuations*. Nobel lecture, 8 December 1977.

[11] Ayala, F.J., Gilpin, M.E. & Ehrenfeld, J.G., Competition between species: Theoretical models and experimental tests. *Theoretical Population Biology.,* **4**, pp. 331–356, 1973.

[12] Beecham, J.A. & Farnsworth, K.D., Animal group forces resulting from predator avoidance and competition minimization. *Journal of Theoretical Biology*, **198**, pp. 533–548, 1999.

[13] Merrifield, A.J., *An Investigation of Mathematical Models for Animal Group Movement, Using Classical and Statistical Approaches.* PhD doctorate, University of Sydney, 2006.

[14] Pykh, Y.A., Entropy characteristics of macrosystems with nonlinear pair interactions, *Doklady Mathematics*, **84**, pp. 669–671, 2011.

[15] Novak, M. & Wootton, T.J., Estimating nonlinear interaction strengths: An observation-based method for species-rich food webs. *Ecology*, **89**, pp. 2083–2089, 2008.

[16] Pykh Y.A., Lyapunov-Meyer functions and distance measure from generalized Fisher's equations. *IFAC-PapersOnLine*, **48–11**, pp. 115–119, 2015.

[17] Pykh, Y.A. & Malkina-Pykh, I.G., Application of the method of double standardization for the development of environmental indices. *Proceedings of the First International Conference on Environmental Indices Systems Analysis Approach (INDEX-97).* EOLSS Publishers: Oxford, pp. 169–181, 1999.

[18] Loitsyanskii, L.G. & Lur'e, A.I., *Course in Theoretic Mechanics*, Nauka: Moscow, 1983.

[19] Pykh, Y.A., Energy Lyapunov function for generalized replicator equations. *Proceedings of International Conference "Physics and Control"*, pp. 271–276, 2003.

[20] Fisher, R.A., *The Genetical Theory of Nature Selection,* Clarendon Press: Oxford, 1930.

[21] Hofbauer, J. & Sigmund, K., Evolutionary game dynamics. *Journal of American Mathematical Society,* **40**, pp. 479–519, 2003.

[22] Pykh, Y.A., Escort distribution, energy and relative entropy of macrosystems with nonlinear pair interactions. *Doklady Mathematics*, **90**, pp. 710–714, 2014.

[23] Pykh, Y.A., *Equilibrium and Stability in the Models of Population Dynamics,* Nauka: Moscow, 1983.

[24] Jensen, G.L.W.V., Sur les fonctions convexes et les ineqalites entre les valenrs moyennes. *Acta Matematica*, **30**(1), pp. 175–193, 1906.

[25] Hardy, G.H., Littlewood, J.E. & Polya, G., *Inequalities*, Cambridge University Press: Cambridge, 1934.

[26] Censor, Y. & Lent, A., An iterative row-action method for interval convex programming. *Journal of Optimization Theory and Applications,* **34**(3), pp. 321–353, 1981.

[27] Bregman, L.M., The relaxation method of finding the common points of convex sets and its application to the solution of problems in convex programming. *Computational Mathematics and Mathematical Physics,* **17**(3), pp. 200–217, 1967.

[28] Legendre, A.M., L'intégration de quelques equations aux differences Partielles. *Memoires de l'Academie Royale des Sciences*, pp. 309–351, 1787.

[29] Donkin, W.F., On a class of differential equations, including those which occur in dynamical problems. *Philosophical Transactions of the Royal Society of London,* **144**, pp. 74–113, 1854.

[30] Fenchel, W., On conjugate convex functions. *Canadian Journal of Mathematics,* **1**, pp. 73–77, 1949.

[31] Amari, S., Differential geometry derived from divergence functions: information geometry approach. *Mathematics of Distances and Applications*, eds M. Deza, M. Petitjean & K. Markov, ITHEA: Sofia, Bulgaria, pp. 9–23, 2012.

[32] Nielsen, F., Legendre transformation and information geometry. *CIG-MEMO,* **1**(2), 2010.
[33] Zia, R.K., Redish, E.F. & McKay, S.R., Making sense of the Legendre transform. *American Journal of Physics,* **77**, pp. 614–625, 2009.
[34] Pykh, Y.A., Construction of entropy characteristics based on Lyapunov energy functions. *Doclady Mathematics*, **69**, 355–358, 2004.
[35] Esteban, M.D. & Morales, D.A., summary of entropy statistics. *Kybernetica,* **31**, pp. 337–346, 1995.
[36] Jaynes, E.T., Information Theory and Statistical Mechanics. *Physical Review*, **106**, pp. 620–630, 1957.
[37] Pykh, Y.A., Lyapunov functions as a measure of biodiversity: Theoretical background. *Ecological Indicators*, **2**, pp. 123–133, 2002.
[38] Pykh, Y.A. & Malkina-Pykh, I.G., Replicator equations, response functions and entropy measures in science: mathematical background. *WIT Transactions on Ecosystems and Sustainable Development V*, WIT Press: Southampton and Boston, pp. 5–15, 2005.
[39] Sobczuk, H. & Lagod, G., Some distance measures for estimating the efficiency of wastewater treatment in activated sludge systems by biomonitoring. *Proceedings of ECOpole,* **5**(2), pp. 431–436, 2011.
[40] Pykh, Y.A., *Pairwise Interactions Origin of Entropy Functions.* Preprint. arXiv: 1506.05731v1 [nlin.AO] 18 June 2015.

Author index

Energy and Sustianability VII

Edited by: ***C.A. BREBBIA****, Wessex Institute, UK and* ***J. J. SENDRA****, Universidad de Sevilla, Spain*

Containing papers presented at the 7th International Conference on Energy and Sustainability, this volume includes collaborative research between different disciplines, including materials, energy networks, new energy resources, storage solutions, waste to energy systems, smart grids and many other related subjects.

Energy production and distribution matters as well as the need to respond to the modern world's dependency on conventional fuels are topics of growing importance. The use of fossil fuels has generated an increasing amount of interest in renewable energy resources and the search for maintainable energy policies.

Energy policies and management are of primary importance to achieve the development of sustainability and need to be consistent with recent advances in energy production and distribution. Challenges lie as much in the conversion from renewable energies such as wind and solar to useful forms like electricity, heat and fuel at an acceptable cost (including environmental damage) as in the integration of these resources into an existing infrastructure.

A range of topics are covered, including: Energy Management; Energy Efficiency; Green Buildings; Energy and Transport; Biomass and Biofuels; Environment and Energy: Connection and Opportunities; Renewable Energy Sources; Waste to Energy; CO_2 Capturing and Management; Case Studies; Processing of Oil and Gas.

WIT Transactions on Ecology and the Environment, Vol 224
ISBN: 978-1-78466-221-9 **eISBN: 978-1-78466-222-6**
Published 2018 / 592pp

www.ingramcontent.com/pod-product-compliance
Lightning Source LLC
LaVergne TN
LVHW062239200926
848484LV00006B/1573
9781784663452